THE VIRAL STORM

THE VIRAL STORM

THE DAWN OF A NEW PANDEMIC AGE

NATHAN WOLFE

TIMES BOOKS HENRY HOLT AND COMPANY NEW YORK

Times Books
Henry Holt and Company, LLC
Publishers since 1866
175 Fifth Avenue
New York, New York 10010

Library of Congress Cataloging-in-Publication Data

Wolfe, Nathan.
The viral storm : the dawn of a new pandemic age / Nathan Wolfe. — 1st ed.
p. cm.
Summary: "The 'Indiana Jones' of virus hunters reveals the complex interactions between humans and viruses, and the threat from viruses that jump from species to species"—Provided by publisher.
ISBN 978-0-8050-9194-6 (hardback)
1. Viruses. 2. Virus diseases. 3. Molecular evolution. 4. Human evolution. I. Title.
QR360.W65 2011
616.9'1—dc22 2011011321

Henry Holt books are available for special promotions and premiums. For details contact: Director, Special Markets.

First Edition 2011

Designed by Kelly S. Too

Printed in the United States of America
1 3 5 7 9 10 8 6 4 2

To my team at Global Viral Forecasting in
San Francisco and around the world who devote their lives
to making the world safe from pandemics

CONTENTS

THE VIRAL STORM

INTRODUCTION

The village of Pang Thruk in the Kanchanaburi province of Thailand is like many in this part of the world—humid, lush, and spilling over with sounds of wildlife. Located in the west of the country near the border of Burma, Pang Thruk is home to about three thousand people whose livelihoods depend on the sugar and rice they grow. In December 2003, it was also home to Kaptan Boonmanuch, a six-year-old boy who would be among the first people to die of a brand-new human virus.

Kaptan loved riding his bicycle, climbing trees, and playing with his plastic toy Dalmatian that pulled three puppies in tiny brown wagons as it barked mechanically. Kaptan also enjoyed helping his family on the farm.

Nearly every family in Pang Thruk kept egg-laying chickens; some also kept roosters for cock fighting. Kaptan's aunt and uncle lived just down the road on their open-air farm of around three hundred chickens. Each winter in the village a few chickens would die from suspected infections or colds, but in December 2003 chicken deaths increased dramatically. That winter, as on many farms in this region, the chickens in Kaptan's uncle's farm suffered from severe diarrhea, strange behavior, and weakness.

All of them either died naturally or were culled as a result of their illness—and Kaptan helped with the dead. A day or two before the New Year, according to reports, the boy carried one of the sick squawking chickens home. That walk home would have lasted no more than a few minutes.

A few days later Kaptan grew feverish. A clinic in the village diagnosed him with a cold, but after three days without improvement, his father, Chamnan, a rice farmer who worked part time as a driver, took him to a public hospital. X-rays revealed that the six-year-old had pneumonia, and he was kept in the hospital for observation. A few days more and Kaptan's fever spiked to a dangerous 105 degrees. His father paid the equivalent of thirty-six dollars for an ambulance to speed him to better care at Siriraj Hospital in Bangkok, more than an hour's drive away.

Upon arrival, Kaptan presented with shortness of breath and fever. Tests revealed that both lungs were affected with severe pneumonia, and the boy was transferred to the pediatric intensive care unit and put on a ventilator. A series of bacterial cultures tested negative, showing that the infection was likely caused by a virus. More detailed testing using a molecular technique called the polymerase chain reaction, or PCR, revealed that Kaptan was likely infected with an atypical type of influenza—perhaps one not yet seen (or seen widely) in humans.

After eleven days of illness, the boy's fever finally began to cool off. However, despite intensive care, his respiratory distress worsened. Just before midnight on January 25, physicians took Kaptan off the respirator. His lungs drowning in fluids, he became Thailand's first known death from H5N1, which would soon become known around the world as "bird flu."

As sad as Kaptan's death was—and the reports go on to describe the boy's funeral and the family's mourning in tragic detail—the reality is that children in the developing world die from diseases like this all the time. And infectious diseases, which scientists in the 1960s predicted would be eliminated in short order, remain some of the most important killers today. But when it comes to global risk, all deaths are not equal. Most deaths from infectious diseases are localized events that, while dire for the victims and their families, present limited risk to the planet as a whole. But some, like Kaptan's, signal a potentially world-altering event: the first human infection by an animal virus that may wipe out millions, or hundreds of millions, of people throughout the planet—permanently changing the face of humanity.

The main objective of my work is to hunt down these events—the first moments at the birth of a new pandemic—and then work to understand and stop them before they reach a global stage. Because pandemics almost always begin with the transmission of an animal microbe to a human, it's work that takes me all around the globe—from rain forest hunting camps of central Africa to wild animal markets of east Asia. But it

Kaptan's brother holds a framed photo of Kaptan Boonmanuch at his funeral. (© *SUKREE SUKPLANG / Reuters / Corbis*)

also takes me to cutting-edge laboratories at the US Centers for Disease Control and Prevention (CDC) and disease outbreak control centers at the World Health Organization (WHO). Tracking down these potentially devastating bugs has led me to study how, where, and why pandemics are born. I work to create systems that can accurately detect pandemics early, determine their likely importance, and, with any luck, crush those that have the potential to devastate us.

As I've lectured on this work around the world and taught undergraduates in my virology seminar at Stanford University, it's hard to ignore a growing general interest in these topics. Everyone recognizes the raw power that pandemics have to sweep through human populations and seemingly kill indiscriminately. Yet, given the importance of these events, large questions remain remarkably opaque:

How do pandemics start?

Why are we now plagued with so many pandemics?

What can we do to prevent pandemics in the future?

This book is my attempt to answer these questions—an effort to assemble the pieces of the pandemic puzzle.

Part I, "Gathering Clouds," introduces our main character, the microbe,* and delves into the history of our relationships with

* Throughout this book, I'll generally refer to *microbes* rather than the more appropriate but clunky term *microorganisms*, which includes all microscopic organisms. Unless otherwise specified, the term *microbe* will be used as shorthand to refer to the full range of microscopic organisms whose groups include species that can infect and spread in humans, namely: viruses, bacteria (and their siblings, the archaea), parasites, and the enigmatic prions, all discussed in detail in chapter 1. While no doubt irritating to some of my microbiologist colleagues, who exclude parasites from the term *microbe* for sensible taxonomic reasons, and haven't quite decided what to do with prions, I hope they will excuse me in this attempt to increase the ease of reading for a general audience.

these organisms. It explores the vast world of microbes, putting those that threaten us in their proper perspective. It details some of the most significant events in the evolution of humans and our ancestors and works to develop the often-spotty historical data into a set of hypotheses on how the events influenced our interactions with microbes.

Part II, "The Tempest," examines how contemporary human populations have grown so exquisitely susceptible to pandemics and what the future years will hold for pandemic diseases. Part III, "The Forecast," describes the fascinating new world of *pandemic prevention* and introduces a new crop of scientists eager to help create a virtual global immune system that will stop pandemics before they become planetary nightmares. Along the way, we will journey to remote hunting villages in central Africa, investigate malaria in wild orangutans in Borneo, learn how cutting-edge genetic sequencing tools will change the way that we discover completely novel viruses, and see how Silicon Valley companies may forever transform the way that we conduct surveillance aimed at finding the next major outbreak.

At this point you may be asking yourself how someone ends up devoting his career to the study of plagues. Is it a desire to save the world? Or perhaps it's the scientific thrill of discovering completely unknown, invisible beings with the potential to wipe out large swaths of humanity. Maybe it's the desire to understand in detail one part of humans' intricate ecology. Or it's an urge to explore the exotic locations on Earth where these novel viruses often appear. But while my life is now consumed with trying to understand and stop pandemics, that's not how it's always been. My work with microbes actually started as a minor footnote to a study I wanted to conduct among wild chimpanzees in central Africa.

As a young child, I acquired what would become a lifelong

interest in apes when I watched a *National Geographic* documentary explaining how humans are more closely related to apes than we are to monkeys. A family tree with humans and apes as siblings (and monkeys as distant cousins) was entirely inconsistent with my memories of seeing these creatures locked up together in the "monkey house" at the Detroit Zoo. We humans were outside the cages, and the rest of *them* were inside. The idea that apes and humans were closely related certainly struck a chord with me. According to my father, I spent some days after the documentary playing the part: walking around the house on all fours, trying to communicate without language, and otherwise working to bring out my inner ape.

My fascination with apes evolved from a childish curiosity to an intellectual interest in what our closest relatives had to tell us about ourselves. What began as a broad interest in apes as animals became a more specific interest in chimpanzees and their less acknowledged brethren, the bonobos—the two ape species that share our own particular branch of the tree of life. How did the years of separation since our last common ancestor with these two kindred ape species shape our minds, our bodies, and our worlds? What features remained the same in all of us?

Along with the intellectual interest, I had a growing desire to see these apes living in their natural environment. This desire required tracking them down in the rain forests of central Africa to see for myself what they were really like. So when choosing among doctoral programs, I settled on one at Harvard where I would work with Richard Wrangham and Marc Hauser, two prominent primatologists. I would spend many months during my first year of doctoral work arguing why they should let me study the troops of wild chimpanzees that Wrangham had worked with for years in the Kibale Forest in southwest Uganda.

I proposed a study to document self-medicating behavior of the Kibale chimpanzees. The idea that these animals consumed plants with specific medicinal chemicals as a way of fighting against their own infectious diseases was still just an hypothesis at the time, and an intriguing one. I'd explored this idea the previous year while studying at Oxford and working on an exhibit about animal self-medication at the Oxford University Museum of Natural History.

The Oxford University Museum is a magnificent nineteenth-century building constructed in the architectural style of a Gothic cathedral but with massive iron supports mimicking the skeleton of a mammal, emphasizing that it was a church of natural history rather than religion. It houses unique collections, including some of the beetles collected by Charles Darwin on his celebrated voyage on the *Beagle*. It had been home to the famous Huxley-Wilberforce debate on natural selection in 1860 seven months after the publication of Darwin's pivotal book *On the Origin of Species*. It's a perfect location to ponder the place of humans in the natural world. The work I did under the supervision of the eminent evolutionary biologist W. D. Hamilton and his colleague

The interior of the Oxford University Museum. (© *Chris Rimmer*)

Dale Clayton, an expert in behaviors animals used to rid themselves of parasites, revealed that self-medication was widespread in the animal kingdom. Animals as disparate as wasps and Kodiak bears utilized the chemical defenses of plants to help rid themselves of their natural pests.

As I began my work in Uganda to study chimpanzees, my professors cautioned me that any convincing proof that chimpanzees were medicating themselves with plants would require an understanding of the infectious diseases they were treating. Unless I could show that the use of the purported medicines decreased the burden of disease, my results would be speculative at best. I needed to understand what infectious diseases plagued the chimpanzees. I knew little about microbes, so I approached Andy Spielman, a professor at Harvard's School of Public Health and one of the few people at the time focused on understanding the ecology of microbes in nature. Despite his lab full of fellows and students and his focus on North America rather than the wilds of Africa or Asia, he kindly took me under his wing. Thus began my research on what was known about the infections of chimpanzees. Once I began thinking about microbes I never looked back. And central to my studies would be the viruses.

Viruses evolve more rapidly than any organism on the planet, yet we understand less about them than any other form of life.* The study of viruses provides a scientist with the

---

* There is some debate about whether viruses themselves should be considered living, while there is no debate about the other microbes: bacteria, archaea, or parasites, all of which are clearly living organisms. The debate, in my opinion, is a semantic and largely unimportant one. Viruses are completely dependent on other organisms for elements of their life cycle, but that is no different than the rest of known life forms, none of which, to my knowledge could live in a world devoid of other life. Either way, it is clear that viruses are part of the living systems of our planet, and for those intent on engaging in this debate, my reference to viruses as living can be interpreted in that way. I will use the same inclusive convention with prions, despite the existence of similar debates on them.

opportunity to discover new species and catalog them in a way reminiscent of the world of the nineteenth-century naturalist, which had so fascinated me during my time at Oxford. A scientist can productively spend an entire career looking for new species of primate and never find one, but new viruses are discovered every year. They also have exceedingly short generations, so we can watch them evolve in real time—an ideal system for someone interested in understanding the process of evolution. Perhaps best of all from the perspective of a young scientist, there was important and urgent low-hanging fruit in this discipline: some of these viruses kill us. Thus new discoveries need not only lead to an improved understanding of nature, but they can also have important and rapid applications for controlling human disease.

Controlling the spread of human disease was at the forefront of public health efforts in early 2004 when the news broke of Kaptan's death from H5N1. His death was the first confirmed mortality from this virus, the so-called bird flu, in Thailand. The truth is that while they may jump to us via other animals, *all* human influenza viruses ultimately originate in birds, so the popular designation of the virus as "bird flu" can irritate scientists. Yet within a month that name would become a mainstay of news shows and a topic of discussion for people throughout the world.

The scientific name for the virus that killed Kaptan, HPAIA (H5N1), is quite descriptive for virologists. It signifies that the virus is a highly pathogenic avian influenza A-type virus and provides the particular hemagglutinin (H) and neuraminidase (N) protein variants particular to this virus strain. But its true significance is actually much more straightforward.

H5N1 is important because it kills remarkably effectively. The virus's case fatality rate, or the percentage of infected individuals that die, is around 60 percent. For a microbe, that's incredibly deadly. As a comparison, we can look back to the devastating 1918 influenza pandemic. While estimates for the 1918 pandemic are imperfect, it is thought around fifty million people died. That's the equivalent of 3 percent of the entire human population at that time, an almost unimaginable catastrophe. To put this in context, more people died from the 1918 influenza pandemic than the total number of soldiers thought to have died in battle during all twentieth-century wars *combined*. More deaths caused by a simple virus, less than one hundred nanometers in diameter and with only a paltry eleven genes, than were caused by all the battles in WWI, WWII, and all of the other wars in our last war-riddled century. Despite the enormity of the 1918 plague, the highest estimates for its case fatality rate are in the range of 20 percent, and it was almost certainly much lower than that; more careful estimates suggest around 2.5 percent.* Recall that H5N1 had a 60 percent fatality rate, far greater than that of the influenza virus that caused the 1918 pandemic.

But while deadliness is an important, dramatic, and ongoing obsession of the media, it is only one piece of the puzzle for microbiologists. In fact, some microbes kill virtually all people they infect: a perfect 100 percent mortality rate. And yet such microbes do not necessarily represent critical threats to humanity. Viruses like rabies, which naturally infect a number of mammal species, and the herpes B virus, which naturally infects some species of Asian monkey, will kill all the people they infect.† Yet unless you're someone who is exposed to rabid

* In fact, the death rate for the 1918 H1N1 infection itself may be even lower than 2.5 percent, as many deaths were probably caused by secondary bacterial infections—deaths that could at least partially be prevented today due to antibiotic use. Deaths from H5N1, on the other hand, are largely due directly to viral disease.

† In the case of rabies, vaccine delivered in short order after infection can successfully prevent death, but without it death is largely inevitable.

animals or works with Asian monkeys, these microbes do not represent a major concern for you. That's because they don't have the capacity to spread from person to person. In order to be catastrophic, a microbe needs both the potential to harm or kill *and* the potential to spread.

In early 2004 there was no way to know how efficiently H5N1 would spread. Since it comes from a class of viruses that often do spread, the influenza viruses, the possibility was there. And if H5N1 were to spread in the same way that the 1918 influenza virus spread, it would create a calamity unlike any seen before in human history.

As impressive as H5N1 is at killing, H1N1, the so-called swine flu,* is equally impressive in its capacity to spread. While no one knows exactly when the H1N1 pandemic began, by August 2009, less than a year after it was first recognized, the WHO announced estimates suggesting that the virus could eventually infect over two billion people, roughly a third of the entire human population. The natural drama of this would be hard to overstate. While less visually dramatic than other forms of natural disaster, the ability of this phenomenon to touch people everywhere on the planet made it a powerful force of nature. A virus that likely infected few people in early 2009 moved around the globe to inhabit a sizable percentage of the entire population of the human species in less than a year. This occurred despite the best efforts of global public health infrastructures—structures that we tend to be very proud of

* Like H5N1, the "swine flu" that began in 2009 suffers from terminology problems. Called H1N1/09 virus by the WHO and 2009 H1N1 influenza among other things by the CDC, here I will refer to it simply as H1N1, which is the commonly used shorthand among the scientists who study it. As with H5N1 and all influenza viruses, H1N1 has its ultimate origins in bird populations.

and feel protected by. Yet while the case fatality rate of H1N1 is estimated to be well below 1 percent, paling in comparison to H5N1, the sheer number of people it infected has made it a real global killer. One percent of two billion is a lot of lives.

To help us understand the real threat of an outbreak, we turn to a concept in epidemiology called R_0, or the *basic reproductive number*. For any epidemic, R_0 is the the average number of subsequent infections that each new case results in (in the context of a population with no prior immunity and no control efforts). If, on average, each case of a new epidemic leads to more than one subsequent infection, the new epidemic has the potential to grow. If, on average, each case leads to less than one subsequent infection, it will peter out. The elegant concept of R_0 helps epidemiologists distinguish between epidemics likely to "go viral" and those likely to go extinct. It's basically a measure of scalability.

Risk interpretation is not a trivial matter, for either the public or the policy makers. In the case of H1N1 or H5N1, the potential costs of not rushing to develop a vaccine or working to decrease transmission could have been global and catastrophic.

Critically, microbes are dynamic—they do not exist in stasis. If H5N1, the deadly bird flu, gains the right combination of genetic mutations it needs to spread effectively, the results will be destructive in a way that, however less visually dramatic, will make even the most deadly earthquakes seem like a walk in the park. And if H1N1, the rapidly spreading swine flu, were to increase in virulence even minimally, its potential to kill would be striking. Neither scenario is implausible. As chapter 1 will explore in detail, influenza and a range of other viruses have an incredible capacity to negotiate the environment of their human hosts. They mutate rapidly, even swapping genes among themselves, a process referred to as reassortment.

It is just such reassortment that concerned me, and other

scientists, in 2009. As the H1N1 virus spread explosively around the world, there was a nontrivial chance that it would run into H5N1 in people or animals, setting the stage for a potentially cataclysmic series of events. These are the kind of events we work to uncover early before they spread. When infected simultaneously with both of these influenza viruses, a particular human or animal could become a potent mixing vessel, providing the perfect opportunity for the bugs to swap genes. How would this happen? In a sort of sexual reproduction, H5N1 and H1N1 could produce mosaic daughter viruses with some of the genes of one virus and some of the genes of the other. Such reassortment events can occur in individuals infected with multiple similar viruses. In the case of H1N1 and H5N1, if that mosaic daughter virus inherited the potential to spread from its H1N1 parent and the deadliness of its H5N1 parent, the resulting virus would be both highly transmissible and highly fatal—exactly the formula for global impact that we most fear.

A mad rush to respond to pandemics has been the mainstay of global public health for the last one hundred years. Now a small but vocal group of scientists and I have begun to argue that we must do better than just *respond* to pandemics by scrambling for vaccines, developing drugs, and modifying behaviors. This traditional approach has proved a failure for human immunodeficiency virus (HIV), which nearly thirty years after its discovery continues to spread, currently infecting over thirty-three million people at last count.

But what if we had been able to catch HIV before it spread? This virus was in humans for over fifty years before it spread widely. It spread for another twenty-five years before French scientists Françoise Barré-Sinoussi and Luc Montagnier, who would go on to win a much-deserved Nobel Prize for their work, finally discovered it. How different would the world be had we stopped it before it left central Africa?

The idea that we might one day predict pandemics is a new one. The first time I heard someone discuss it was around ten years ago in the Johns Hopkins office of Don Burke, a retired medical colonel and world-renowned virologist from the Walter Reed Army Institute of Research (WRAIR) who had devoted his life to more traditional approaches to control disease before he took on a professorship at Johns Hopkins University's Bloomberg School of Public Health. Don had hired me as a postdoctoral fellow at Hopkins a few years earlier when I was finishing up my doctoral work in the rain forests of north Borneo studying the ways that mosquitoes and other blood-feeding insects helped microbes move between primate species.

Unable to track me down himself, Don had managed to find my mother in Michigan and gave her a call. My mother, whom I would speak to infrequently on trips from our forest research station, scolded me, saying that a "general" from the US military had called her. She asked what kind of trouble I had managed to get myself into. Fortunately, all Don wanted was to see if I'd help him set up a project in central Africa to understand how viruses emerge from animals into human populations.

During the years that followed, in addition to the long, slogging work of building research capacity to catch new microbes in central Africa and Asia, Don and I engaged in hours of conversation in the field and in his office in Baltimore. We made numerous beer bets on scientific problems and asked hard questions about the future of our field. I remember the day when I first heard Don suggest that the future would include not only *response* to pandemics but *prediction*. It seemed a bold but logical idea, and we quickly moved toward thinking about ways that it could actually happen. Those early conversations formed the foundation of the work my colleagues and I conduct now, setting up and running listening posts at microbial hot spots around the world aimed at catching new microbes locally before they become global pandemics.

Among the things we listen for are novel influenza viruses, like H5N1 and H1N1. Unfortunately, the world too easily becomes complacent to threats like H5N1 and H1N1. These and other threats fade quickly from the media's attention. Most of the world doesn't think about either of these viruses seriously. Yet neither virus has gone extinct, and the threat is perhaps as great now as it was when they were each first noticed. Both viruses continue to infect human populations. For example, in 2009, years after the media forgot about the virus, H5N1 caused at least seventy-three laboratory-confirmed cases. This is almost certainly a substantial underestimate of the number of actual cases and does not differ notably from the number of confirmed cases from previous years. H1N1 cases continue to spread as well. We have detected them even in the most remote forest areas where we "listen."

Amazingly, in a time when we can sequence an entire human genome for under ten thousand dollars and build massive telecommunications infrastructures that will soon make cell phones available to most people on the planet, we still understand surprisingly little about pandemics and the microbes that cause them. We know even less about how to predict or prevent pandemics before they spread from small towns to cities and the rest of the world. As I'll argue in part II, pandemics will increase in frequency in the coming years as the connections between human populations and the animals in our world continue to grow. Whether it's a mosaic virus with the deadliness of H5N1 and the potential to spread like H1N1, a resurgent SARS, a new retrovirus like HIV, or perhaps most frighteningly, a completely novel microbe that blindsides us, microbial threats will grow in the coming years in their ability to plague us, kill people, destroy regional economies, and threaten humanity in ways more severe than the worst imaginable volcanoes, hurricanes, or earthquakes.

A storm is brewing. The objective of this book is to understand this coming storm—to explore the nature of pandemics, to understand where they come from and where they are going. But it will not paint a completely grim picture. In the one hundred years since we first discovered viruses, humans have come a long way in understanding them. Much hard work remains. Yet if we do it well, we will harness the numerous technological advances of our time that provide tools to predict pandemics—just as meteorologists predict the course of hurricanes—and ideally prevent them from occurring in the first place. This is the Holy Grail of modern public health, and in the coming pages, I will argue that it is within our grasp.

PART I

GATHERING CLOUDS

► 1 ◄

THE VIRAL PLANET

Martinus Beijerinck was a serious man. In one of the few images that remain of him, he sits in his Delft laboratory in the Netherlands, circa 1921, just a few days before his reluctant retirement. Bespectacled and in a suit, he's presumably as he'd like to be remembered—among his microscopes, filters, and bottles of laboratory reagents. Beijerinck had some peculiar beliefs, including the idea that marriage and science were incompatible. According to at least one account, he was verbally abusive to his students. While rarely remembered in the history of biology, this strange and serious man conducted the pivotal studies that first uncovered the most diverse forms of life on Earth.

Among the things that fascinated Beijerinck in the late nineteenth century was a disease that stunted the growth of tobacco plants. Beijerinck was the youngest child of Derk Beijerinck, a tobacco dealer who went bankrupt due to crop losses caused by this blight. Tobacco mosaic disease causes discoloration in young tobacco plants, leading to a unique mosaic pattern on leaves and radically slowing the growth of the adult plants. As a microbiologist, Beijerinck must have been frustrated by the

Dr. Martinus Beijerinck.
(*Undated photo*)

unclear etiology of the disease that had wiped out his father's business. Despite the fact that it spread like other infections, microscopic analysis did not reveal a bacterial cause of disease. Curious, Beijerinck subjected the fluids of one of the diseased plants to intense filtration using a fine-grained porcelain filter. He then demonstrated that even after such filtration the fluids retained their capacity to infect healthy plants. The tiny size of the filter meant that bacteria, the usual suspects for transmissible disease at the time, would be too large to pass through. Something else must have caused the infection—something unknown and considerably smaller than everything else recognized to be alive in his time.

Unlike his colleagues, many of whom believed a bacteria would emerge as the cause, Beijerinck concluded that a new form of life must cause tobacco mosaic disease.* He named this

* There are some who consider Dmitri Ivanovski the "father of virology" because he did similar research with tobacco mosaic virus six years earlier. But perhaps because he wasn't the first to name the new entities (i.e., viruses) or did not as widely disseminate his findings as Beijerinck, he is not generally credited with their discovery.

new organism the *virus*, a Latin word referring to poison. The word *virus* had been around since the fourteenth century, but his use was the first to link it to the microbes to which it refers today.* Interestingly, Beijerinck referred to viruses as "contagium vivium fluidum," or "soluble living agent," and felt they were likely fluid in nature. That is why he used the term *virus*—or poison—to denote its "fluidity." It wasn't until later work with the polio and foot-and-mouth-disease viruses that the particulate nature of viruses was confirmed.

In Beijerinck's time a new microscopic perspective began revealing itself to scientists. Looking through microscopes and applying gradually smaller filters, these microbiologists realized something that continues to amaze us today: shielded from our human-scaled senses is a wide, teeming, startlingly diverse, unseen world of microbial life.

I teach a seminar at Stanford called Viral Lifestyles. The title was meant to evoke curiosity among prospective students but also describe one of the course's objectives: to learn to envision the world from the perspective of a virus. In order to understand viruses and other microbes, including how they cause pandemics, we need to first understand them on their own terms.

The thought experiment that I give my students on the first day is this: imagine that you have powerful glasses allowing you to perceive any and all microbes. If you were to put on such magical bug-vision specs, you would instantly see a whole

* In addition to his pivotal work as the first virus hunter, creating the foundations of what would later become the field of virology, Beijerinck remains an unsung hero for those studying the relationships between plants and bacteria. Among other notable findings, he discovered nitrogen fixation, whereby bacteria living in the roots of legumes make nitrogen available to plants through a set of biochemical reactions critical for the fertility of agricultural soil systems.

new, and very active, world. The floor would seethe, the walls would throb, and everything would swarm with formerly invisible life. Tiny bugs would blanket every surface—your coffee cup, the pages of the book on your lap, your actual lap. The larger bacteria would themselves teem with still smaller bugs.

This alien army is everywhere, and some of its most powerful soldiers are its smallest. These smallest of bugs have integrated themselves, quite literally, into every stitch of the fabric of earthly life. They are everywhere, unavoidable, infecting every species of bacterium, every plant, fungus, and animal that makes up our world. They are the same form of life that Beijerinck found in the late days of the nineteenth century, and they are among the most important of the microbial world. They are viruses.

Viruses consist of two basic components, their genetic material—either RNA or DNA—and a protein coat that protects their genes. Because viruses don't have the mechanisms to grow or reproduce on their own, they are dependent on the cells they infect. In fact, viruses *must* infect cell-based life forms in order to survive. Viruses infect their host cells, whether they are bacterial or human, through the use of a biological lock-and-key system. The protein coat of each virus includes molecular "keys" that match a molecular "lock" (actually called a receptor) on the wall of a targeted host cell. Once the virus's key finds a matching cellular lock, the door to that cell's machinery is opened. The virus then hijacks the machinery of that host cell to grow and propagate itself.

Viruses are also the smallest known microbes. If a human were blown up to the size of a stadium, a typical bacterium would be the size of a soccer ball on the field. A typical virus would be the size of one of the soccer ball's hexagonal patches. Though humans have always felt virus's effects, it's no wonder it took us so long to find them.

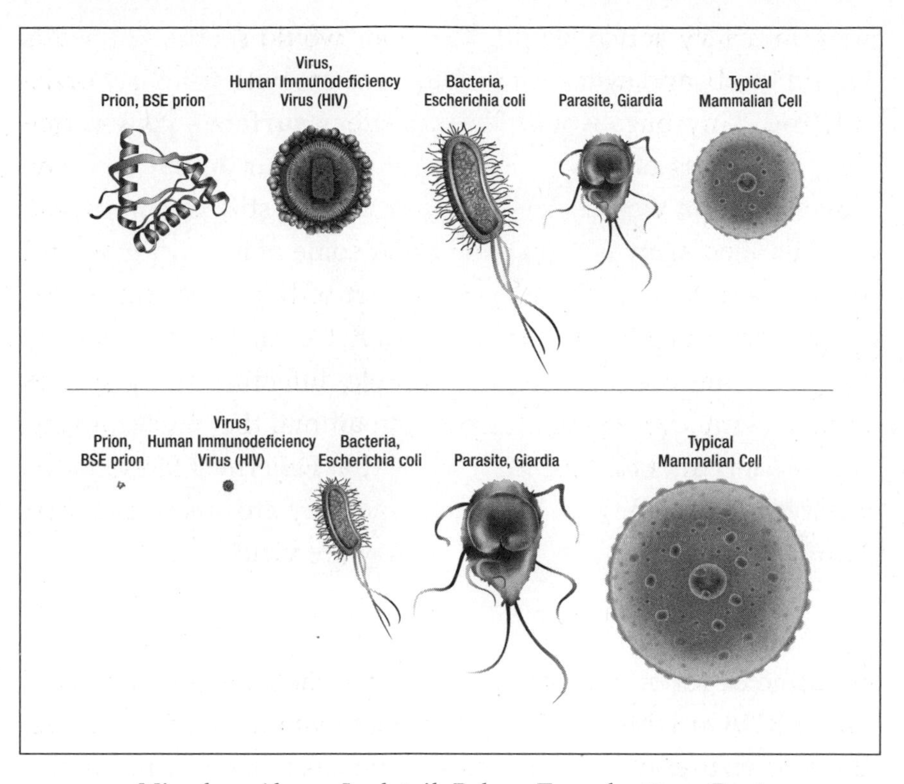

Microbes, Above: In detail; Below: To scale. (*Dusty Deyo*)

Viruses, the most diverse forms of life, remained completely opaque to humans until a meager one hundred years ago with Beijerinck's discovery. Our very first glimpses of bacteria came a little under four hundred years ago when Antonie van Leeuwenhoek adapted the looking glasses of textile merchants to create the first microscope. With it, he saw bacteria for the first time. This finding represented such an incredible paradigm shift that it took the British Royal Society another four years before it would accept that the unseen life forms were not merely artifacts of his unique apparatus.

Our scientific understanding of unseen life has proceeded pitifully slowly. Compared to some of the other major scientific breakthroughs over the last few thousand years, our understanding of the dominance of unseen life occurred only recently. By

L: Replica of van Leeuwenhoek's microscope, 17th century; R: van Leeuwenhoek's microscope in use. (*L: Dave King / Getty Images; R: Yale Joel / Getty Images*)

the time of Jesus, for example, we already understood critical elements of how the Earth rotated, its rough size, and its approximate distance to the sun and moon—all fairly advanced elements in understanding our place in the universe. By 1610 Galileo had already made his first observations using a telescope. Van Leeuwenhoek's microscope came fifty years after that.

It is hard to overstate the paradigm shift that van Leeuwenhoek's discovery represents. For thousands of years humans had recognized the existence of planets and stars. Yet our understanding of unseen life and its ubiquity began only a few hundred years ago with the invention of the microscope. The discovery of novel life forms continues to this day. The most recent novel life form to be uncovered is the unusual prion, whose discovery was acknowledged with a Nobel Prize in 1997. Prions are an odd microscopic breed that lack not only cells but also DNA or RNA, the genetic material that all other known forms of life on Earth use as their blueprint. Yet prions persist and can be spread, causing, among other things, mad cow disease. We would be arrogant to assume that there are no other life forms remaining to be discovered here on Earth, and they are most likely to be members of the unseen world.*

We can roughly divide known life on Earth into two groups: *noncellular* life and *cellular* life. The major known players in the noncellular game are viruses. The dominant cellular life forms on Earth are the prokaryotes, which include bacteria and their cousins, the archaea. These life forms have lived for at least 3.5 billion years. They have striking diversity and together make up a much larger percentage of the planet's biomass than the other more recognizable cellular forms of life, the eukaryotes, which include the familiar fungi, plants, and animals.

Another way of categorizing life is this: seen and unseen. Because our senses detect only the relatively large things on Earth, we are parochial in the way that we think about the richness of life. In fact, unseen life—which combines the worlds of bacteria, archaea, and viruses as well as a number of microscopic eukaryotes—is the truly dominant life on our planet. If some highly advanced extraterrestrial species were to land on Earth and put together an encyclopedia of life based on which things made up most of Earth's diversity and biomass, the majority of it would be devoted to the unseen world. Only a few slender volumes would be dedicated to the things we normally equate with life: fungi, plants, and animals. For better or worse, humans would make up no more than a footnote in the animal volume—an interesting footnote but a footnote at best.

Global exploration to chart the diversity of microbes on the

* Among the most intriguing possibilities is that non-DNA/non-RNA forms of life, which originated completely independently of our own RNA/DNA-based life, might persist undetected on Earth. These life forms, referred to as shadow life, would almost certainly be microscopic. If discovered, they might best be described as aliens, and some believe that if we are to discover aliens within our lifetimes, looking on Earth will be our best shot.

planet remains in its infancy. Considering viruses alone gives some sense of the scale of what's unknown. It's thought that every form of cellular life hosts at least one type of virus. Essentially—if it has cells, it can have viruses. Every alga, bacterium, plant, insect, mammal. Everything. Viruses inhabit an entire microscopic universe.

Even if every species of cellular life harbored only one unique virus, that would by definition make viruses the most diverse known life forms on the planet. And many cellular life forms, including humans, harbor a range of distinct viruses. They are found everywhere—in our oceans, on land, deep underground.

The dominant forms of life on our planet, when measured in terms of diversity, are unambiguously microscopic.

The largest virus to be discovered is the still microscopic six-hundred-nanometer Mimivirus—viruses are by nature tiny. But the sheer number of viruses in our world leaves a significant biological impression. A groundbreaking paper published in 1989 by Oivind Bergh and his colleagues at the University of Bergen in Norway found up to 250 million virus particles per milliliter of seawater, using electron microscopy to count the viruses. Alternate, more comprehensive measurements of the biomass of viruses on Earth are even more unimaginably outsized. One estimate suggests that if all the viruses on Earth were lined up head to tail the resulting chain would extend 200 million light years, far beyond the edge of the Milky Way. Though often thought of as a pesky irritant or blight, viruses actually serve a role that goes far beyond, and has a much greater impact than, what was previously understood—a role that scientists are only just beginning to comprehend.

It's true that in order to complete their life cycle, viruses have to infect cellular forms of life, but their role is not necessarily destructive or harmful. Like any major component of the global ecosystem, viruses play a vital role in maintaining global equilibrium. The 20 to 40 percent of bacteria in marine ecosystems that viruses kill every day, for example, serves a vital function in the resulting release of organic matter, in the form of amino acids, carbon, and nitrogen. And though studies in this area are few, it is largely believed that viruses, in any given ecosystem, play the role of "trust busters"—helping to ensure that no one bacterial species becomes too dominant—thereby facilitating diversity.

Given the ubiquity of viruses, it would be surprising indeed if they were relegated to a destructive role. Further studies will likely reveal the profound ecological importance of these organisms not just in destroying but also in benefiting many of the life forms they infect. Since Beijerinck's discovery, the vast majority of research conducted on viruses has understandably focused on the deadly ones. In the same way, we know much more about venomous snakes, despite the fact that they represent a startlingly small percentage of snake diversity. As we consider the frontiers of virology in part III, we will explore the potential benefits of viruses in detail.

Viruses infect all known groups of cellular life. Whether a bacterium living in the high-pressure depths of the planet's upper crust or a cell in a human liver, for a virus, each is just a place to live and produce offspring. From the perspective of viruses and other microbes, our bodies are habitats. Just as a forest provides a habitat for birds and squirrels, our bodies provide the local environment in which these beings live. And survival in these environments presents a range of challenges. Like all forms of life, viruses compete with each other for access to resources.

Viruses face constant pressure from our immune systems, which have multiple tactics to block their entrance into the body or disarm and kill them when they manage to get in. They face constant life choices: should they spread, which risks capture by our immune systems, or remain in latency, a form of viral hibernation, which can provide protection but sacrifices offspring.

The common cold sore, caused by the herpes simplex virus, illustrates some of the challenges that viruses face in negotiating the complex habitats of our bodies. These viruses find refuge in nerve cells, which because of their privileged and protected positions in our bodies do not receive the same level of immune attention as the cells in our skin, mouth, or digestive tract. Yet a herpes virus that maintained itself within a nerve cell without spreading would hit a dead end. So herpes viruses sometimes spread down through the nerve cell ganglions to the face to create virus-loaded cold sores that provide them a route to spread from one person to the next.

How viruses choose when to launch themselves remains largely unknown, but they almost certainly monitor the environmental variables of their world when making these decisions. Many of the adult humans who are infected with herpes simplex virus know that stress can bring on cold sores. Some also have noted anecdotally that pregnancy seems to bring on active infections. While still speculation, it would not be surprising if viruses responded to environmental cues indicating severe stress or pregnancy by activating. Since severe stress can indicate the possibility of death, it may be their last opportunity to spread—a dead host is also a dead virus. A pregnancy, on the other hand, presents the opportunity for spread either through genital contact with the baby during childbirth or during the kissing that inevitably follows the birth of a baby.

Transmission from host to host is such a fundamental need for infectious agents that some take it a step further. The incredible malaria parasite *Plasmodium vivax hibernans* goes so far as to keep a calendar of sorts. Many times larger than herpes simplex virus, parasites like malaria are infectious agents like viruses and bacteria but are in the eukaryotes class, and so are more closely related to animals than they are to the others. Spread by mosquitoes, *P. vivax hibernans* persists in arctic climates. In these cold locations, it can only infect mosquitoes seasonally during the brief period each summer when the insects hatch. Rather than wasting energy producing offspring all year, the malaria parasite lies dormant for most of the year in the human liver but, during summer, bursts to life, generating its spawn of malaria offspring that spread through an infected person's blood. While it's still unclear exactly what triggers the relapse, recent studies suggest that it might be the bites of mosquitoes themselves that provide an indication that the season for spreading has begun.

The careful timing that viruses and other microbes use when choosing to spread does not differ from the choices that other organisms make. Whether the timing of fruiting in a tropical fruit tree or the timing of mating in water buffalo, living things that time their reproduction appropriately have more successful offspring. This means the traits for accurately timing reproduction will persist and diversify. And how microbes time their growth within our bodies also has a major impact on illness.

The majority of microbes that cause infection in humans are relatively harmless, but some have a striking capacity to make us sick. This can sometimes be expressed in the form of, say, a common cold (caused by a rhinovirus or adenovirus) but can also manifest itself in life-threating illnesses such as smallpox.

Deadly microbes are a consistent challenge to evolutionary biologists because of their paradoxical habit of eviscerating habitats upon which they depend for their own survival. It's analogous to a bird destroying the forest in which it and its descendants live. Yet the process of evolution occurs largely at the level of the individual or even the gene. Evolution does not proceed with forethought, and there's nothing to stop a virus from spreading in such a way that leads to a dead end. Such virally induced extinction events have undoubtedly occurred throughout the history of interactions with microbes, no matter the ultimate cost for virus or host.

More central from the perspective of a virus is the impact of disease on transmission. As we learned in the introduction, on average, each germ must infect at least one new victim for every old one who either dies or recovers and purges himself of the microbe in order to avoid extinction. This is the rule of the basic reproductive number, or R_0. If the average number of new victims per old victims drops to less than one, then the spread of the microbe is doomed. Since microbes generally can't walk or fly from one host to the next they often strategically alter their host to help in their spread. From the perspective of a bug, a symptom can be an all-important means of enlisting our help in moving itself around. Microbes often make us cough or sneeze, which can permit them to spread through our exhaled breath, suffer from diarrhea, which can spread microbes through local water supplies, or cause open sores to appear on our skin, which can spread through skin-to-skin contact. In these cases it's obvious why a microbe would trigger these generally unpleasant symptoms. Unpleasant symptoms are one thing, but killer microbes are quite another.

Keeping its host alive and pumping out new microbes would seem to be an ideal plan for a bug. And some bugs do certainly employ such a strategy. Human papillomavirus, or HPV, infects around 50 percent of sexually active adults at some point during their lifetimes. It currently infects around 10 percent of people on the planet, a staggering 650 million people. And while a few strains of HPV cause cervical cancer, most do not. Those strains that do kill their hosts infect them for many years before showing any symptoms at all. Even if the current vaccines that protect against the cancer-causing HPV variants were deployed universally, harmless HPV strains would continue to circulate at huge levels with an impact no greater than occasional if unsightly warts. These viruses can spread effectively without killing. Yet other bugs kill with startling efficiency.

Bacillus anthracis, a bacterial pathogen of grazing animals like sheep and cattle that occasionally infects humans, causes anthrax infection, which kills quickly and effectively. Following ingestion of anthrax spores during grazing, anthrax reactivates and spreads rapidly throughout the animal, often killing it in short order. But this dead host is by no means a dead end. After using the energetic resources of its dying host to replicate in massive numbers, anthrax simply goes back into spore form. Wind, a common feature of the grassy plains of the grazing hosts, then spreads the spores throughout the environment, where they can wait for new prospective victims to arrive. In the case of anthrax, creating hardy spores frees the bug from any negative consequences of its destruction.

Such situations are not limited to spore-forming bacteria. The cholera bacterium, which gives us diarrhea, and the smallpox virus, which causes severe viral disease, both kill in only days or weeks. But before the deaths take place, the deadly symptoms spread trillions of microbes to potential new victims. Human deaths, while unfortunate to us, represent a mere consequence of the conditions the bugs need to get to their next hosts.

From the perspective of a bug the impact on its host is only measured in its ability to survive and reproduce. And altering our physical bodies is just the beginning. Some microbes also influence our behavior, effectively making us zombies acting in their benefit. One of the most striking examples comes from a feline parasite, *Toxoplasma gondii.* While toxo, as parasitologists refer to it, can infect a range of mammals from humans to rodents, its life cycle cannot be completed until it lands in a cat. This parasite has found a frighteningly effective way to get home when it ends up in the wrong mammal. Careful studies have documented how the parasite spreads to the nervous system of infected, unsuspecting rodents and hijacks their brains. Sometime after infection, having spent much of their life steering clear of cats, mice begin to see them as positively enticing. This fatal attraction leads to a dead mouse, but also a toxo cyst that has the potential to complete its life cycle in the newly infected, not to mention, satiated, cat.

Truly deadly diseases must strike a balance between the likelihood of causing death in its victim once the victim is infected and their efficacy in terms of allowing the victim to spread the disease to others. You can't generally have your cake and eat it, too—producing many microbes in a host increases the chance that they'll spread but also harms the host. Consequently, microbes sometimes use very different methods to cause devastation. They can keep the carrier alive for a long time, which carries the potential to infect multiple victims over many months or years, as in the case of the HPV virus. Or they can kill and spread quickly, infecting dozens of new victims in the course of a day, as in the case of smallpox and cholera.

That a tiny microbe has the potential to alter the body and behavior of its host represents an enormous logistical feat. As scientists sequence the genomes of different species, they provide information on the relative size of the genetic blueprints that permit these organisms to function and give us a sense of how enormous the feat is. The numerical genome sizes of many cellular forms of life can range into the billions—humans, for example, have around three billion base pairs (i.e., bits of genetic information); corn has around two billion. Certain viruses like HIV and the Ebola virus, which use RNA rather than DNA for their genetic information, manage to live with an average of only ten thousand base pairs of genetic information, an incredible level of biological minimalism. How they manage to replicate with such a small amount of genetic information, let alone do something remarkably complicated, like altering the behavior of their hosts, is truly amazing.

Viruses manage to function with such few genes through a variety of tricks that allow them to maximize the impact of their diminutive genomes. Among the most elegant is a phenomenon called *overlapping reading frames.* As an analogy, take a poem of around thirteen thousand letters—say, T. S. Eliot's poem *The Waste Land.* It has roughly the same number of letters as the Ebola virus has base pairs. When you read *The Waste Land,* it has meaning, tempo, reference—all of the characteristics we normally expect from literature. In the same way, the genome of the Ebola virus has meaning, with base pair letters making up genes that get translated into the proteins that provide the virus with its capacity to function. If you take the first stanza of *The Waste Land,* around a thousand letters, and begin to read it starting with the second letter instead and move the first letters of the other words, it's a disaster. "April is the cruelest month" becomes "Prili sthec rueles tmonth." Nonsense.

Now imagine that embedded within the stanza was a second poem so that both readings, the one that starts with the first letter and the one that starts with the second letter, lead to fluent comprehensible verses. Now imagine that you took the same stanza and read it backward and that a third hidden stanza emerged from the same letters. This is precisely what viruses can do. A good challenge to poets (or perhaps computer scientists) would be to create such a stanza to see if they could be as creative as natural selection has been with viruses.

Viruses with overlapping reading frames use the same string of base pairs to code up to three different proteins, an incredible genomic efficiency, which makes their small genomes pack a much larger punch.

Overlapping reading frames represent just one of a range of adaptations that viruses have to negotiate their worlds. Perhaps even more important for viruses is their capacity to generate genetic novelty. Viruses have a diverse toolbox for altering themselves. Among the most fundamental is simple mutation. No organisms have perfect fidelity. Any time a cell in our body or a bacterium divides to create daughter cells or a virus replicates in a host cell, errors creep in. This means that even in the absence of sexual mixing, offspring are never the same as their parents. Yet viruses have taken mutation to a completely new level.

Viruses have some of the highest mutation rates of any known organisms. Some groups of viruses, such as RNA viruses, have such high error rates that they approach a threshold where any higher level of mutation would make them effectively crash due to the loss of essential function from the resulting errors. While many of the mutations harm the new viruses, the high number of offspring that viruses produce increases the chances that some mutants survive and occasionally outperform their parents. This raises the chances that they will successfully evade the immune systems of their host, get the upper hand against a new drug, or gain the capacity to jump to a completely new host species.

Middle-school biology teaches us that life is made up of sexual or asexual organisms. Yet viruses and other microbes exchange genetic information in ways that should make us question our early textbooks. When two different varieties of virus infect the same host, from time to time they infect the same cell, setting the stage for such exchange. In these cases, viruses sometimes create mosaic daughter viruses, which include some genetic parts from one of the viruses and completely different elements from another. In the case of reassortment, entire gene segments are swapped between certain kinds of viruses. In recombination, genetic material from one virus is swapped into a second virus. Genetic mixing of both sorts provides viruses with a rapid and radical way to create novelty. As with mutation, the novel daughter viruses have new blueprints that occasionally help them survive and spread.

Our knowledge of microbes is still young. This vast unseen world is critical to our planet and our species, yet we understand very little about it. We've already discovered most of the plant and animal life on our planet, but we regularly discover brand-new microbes. Ongoing studies of the diversity of microbes in animals, plants, soils, and aquatic systems represent the tip of a very large iceberg. The millions of specimens that will result from these studies will catalyze our understanding of life. Among other things, the knowledge will help spark the development of new antibiotics. It will also help us forecast the next pandemic. The microbial world is the "new world," the last frontier of undiscovered life on our planet.

➤ 2 ➤

THE HUNTING APE

I wiped the sweat out of my eyes and swatted away the prickly branches in my path as I tried to listen for the screeches and hollers of the wild chimpanzees my colleagues and I had been trailing through Uganda's Kibale Forest for the past five hours. The sudden silence of the three large male chimpanzees could only mean trouble. At times, such silence can foreshadow a sudden murderous rush into a neighboring territory to kill competing males. Or perhaps scientists. Chimpanzee warfare was not, thankfully, in the air that day. When our group emerged into a small clearing, we observed the chimpanzees seeming to quietly confer with one another as a crew of red colobus monkeys ate and played in the fig trees above, unaware of any danger. As two of the males inched up two nearby trees, the third—the apparent leader—created a diversion by screaming and scrambling up the tree toward the monkeys. Commotion ensued as the monkeys scrambled out of the tree and landed in the path of the other two hunters, waiting. One of the chimpanzees grabbed a young monkey and made his way to the ground to share his catch with his teammates.

As the chimpanzees feasted on the monkey's raw flesh, a

rush of thoughts ran through my brain: teamwork, strategy, flexibility. All in this close relative to humans. Truly, this was why people studied chimpanzees. While the rigors of scientific literature would never allow us to state this in technical journal articles, the reality seemed clear enough—these chimpanzees had worked collectively and strategically to mount a coordinated attack. The leader had diminished his chances of landing a kill by making a noisy attack, but the knowledge that his actions would increase the chances of success for his partners made this a strategic approach. In the end, they'd share the meat no matter who made the kill, exactly the sort of behavior that humans display every day. As the chimpanzees tore through the animal, it also occurred to me that the contact with the monkey's blood and guts provided the ideal opportunity for our carnivorous kin to contract microbes.

Studying our closest living primate relatives affords us the opportunity to better understand ourselves, genetically, socially, and otherwise. However imperfect the conclusions we draw about ourselves from studying wild primates, we're lucky to have them since the fossil record only offers its gems sporadically. Humans love the idea that we're the chosen species—unique among the members of the animal kingdom—yet such claims should meet a high standard of proof. If our ape cousins share our supposedly unique traits, then perhaps they're not unique traits after all. If, for example, we'd like to know if humans evolved the capacity to hunt or share food independently, we can look to chimpanzees and bonobos and ask if they exhibit the same behaviors. If they do, then Occam's razor should push us toward concluding that we all share these traits because of shared descent: evolving the ability to hunt collectively twice or thrice within the very same close lineage is a less parsimonious explanation than simply concluding that hunting emerged in our joint ancestors before we split with them.* That a human trait is interesting does not mean it is unique to us. Many undoubtedly have ancient origins.

Some people have an almost instinctually negative response to the discovery that a treasured aspect of humanity is in fact not unique—that it's actually something we share with other animals. Of course, the objective of science is not to uncover the things that make us comfortable but rather the things as they are. Another perspective on these shared traits is that they can help us feel less alone and more connected to the rest of life on our planet.

The parsimony rule of thumb applies not only to our behaviors. Each organ, each cell type, each infectious disease presents a new point of comparison with our kin. Are they found in us alone, or are they found in multiple other species along our same branch of the evolutionary tree? Through careful studies of humans and our closest living relatives we have the potential to at least begin to sort through historical mysteries and solidify which elements of humanity are unique and which are not. Already, earlier ideas that human traits like using tools or fighting wars were unique have been overturned by discoveries that chimpanzees engage in the same behaviors. What other supposedly unique human traits will fall next remains to be seen.

Fortunately, we have close living relatives that we can observe. The apes, our own branch of the primate lineage, include humans, chimpanzees, bonobos, as well as gorillas, orangutans, and the

* Sadly, using actual fossil evidence, such as tooth wear and carbon typing methods, to address these questions remains imperfect. They indicate that just as for chimpanzees and bonobos, the majority of food for our ancestors prior to around 1.8 million years ago was of plant origin. But meat was almost certainly a part of the diet—tool-scarred bones have been found that are over three million years old, and tooth wear patterns indicate heavy meat eating by around two million years ago.

least studied apes, the gibbons. Studies of ape skeletons during the past hundred years provide a rough guide to the historical relationships among all of us. Over the last decade, a mass of genetic data from these animals has further refined the picture, providing a clear pattern of primate relationships. The information, commonly represented by the geneticists who study these data in phylogenetic trees such as the one below, helps to graphically describe how the relationships shake out.

The research reveals that for humans, two key species, chimpanzees and bonobos, lie closest to us. The other apes (gorillas, orangutans, and gibbons) differ substantially more and thus represent distant cousins of our human-chimpanzee-bonobo group. This relationship has led to the notion that humans are best seen as *the third chimpanzee* species, described in great detail in Jared Diamond's book of the same title.

Once referred to as pygmy chimpanzees, scientists now recognize bonobos as an entirely separate species, yet one closely related to chimpanzees. Bonobos live only south of the Congo

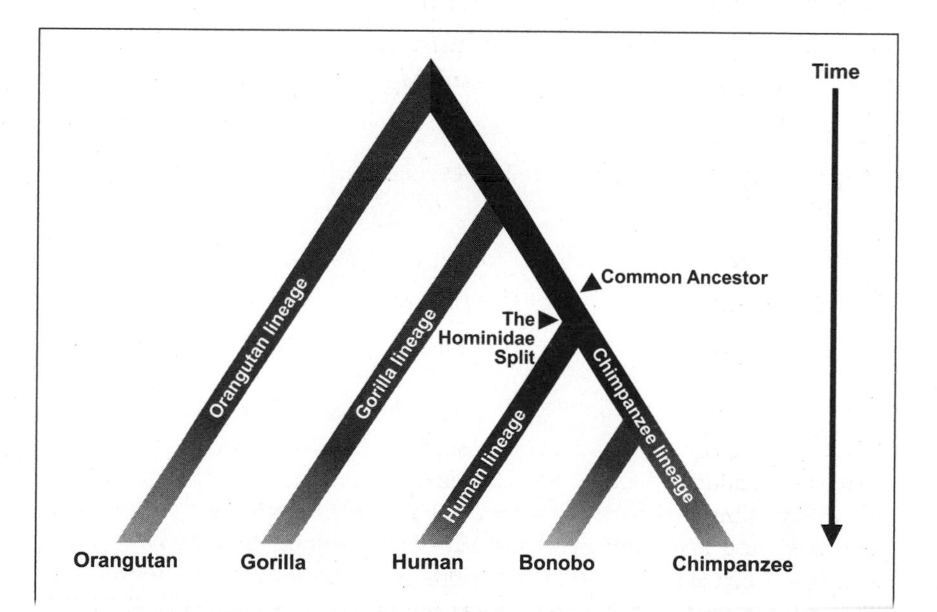

Phylogenetic tree, representing the evolution of apes. (*Dusty Deyo*)

River in central Africa, while chimpanzees live only north of it. And while they look very similar, bonobos and chimpanzees have evolved to exhibit significant differences in their behavior and physiology during the time they've been separated by the great river. Current estimates suggest that the chimpanzees and bonobo lineages diverged roughly one to two million years ago. This divergence occurred some time after our own lineage separated from these cousins, around five to seven million years ago.

This research helps point us to a very pivotal and informative character in the evolution of our own species, a character referred to by anthropologists as the most recent common ancestor, which I'll refer to simply as the *common ancestor.* Around eight million years ago in central Africa lived an ape species whose descendants would go on to include humans as well as the chimpanzees and bonobos.

We can use our parsimony rule of thumb and simple common sense to imagine the common ancestor in a bit more detail. It had extensive body hair and likely spent much of its time in the trees as do chimpanzees and bonobos. It lived in

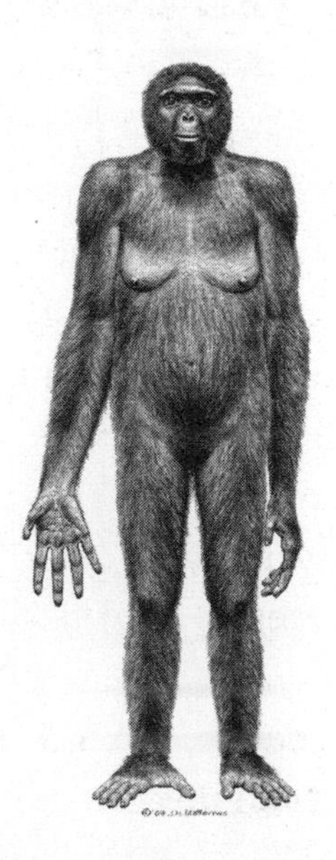

An artist's conception of "Ardi," a female *Ardipithecus ramidus*, 4.4 million years old, representative of the most recent common ancestor between humans and chimpanzees. (*Science Magazine / Jay Matternes*)

central Africa and consumed a diet dominated by fruit, tropical fruit in the fig family probably making up the major staple. Had we been able to study this ape, it would certainly have told us important things about what would come for us in the future, what changes were brewing. One thing that would end up affecting the future of our relationship with infectious diseases was a new tendency present in this animal: the urge *and* ability to hunt and eat meat.

That humans share with chimpanzees the trait of hunting animals has been known for some time. It first emerged in the early 1960s when the British primatologist Jane Goodall documented wild chimpanzees hunting and eating meat at Gombe National Park in Tanzania during her pioneering efforts to study wild chimpanzee behavior. Before the Goodall studies and a related set of studies conducted by Japanese colleagues in the Mahale region of Tanzania, our understanding of chimpanzee behavior in the wild was largely nonexistent. The finding that chimpanzees hunted came as a shock to anthropologists, many of whom had come to believe that hunting had emerged after our split with chimpanzees and shaped our evolution in a way that distinguished us from them.

Since then, detailed studies in Gombe and Mahale as well as in some of the half-dozen more recently studied wild chimpanzee communities have solidified our understanding of the important role of meat in the chimpanzee diet. While chimpanzees hunt opportunistically, it is by no means sporadic. Chimpanzees can hunt forest antelopes and other apes (even humans), but they tend to specialize in a few critical species of monkey as prey. Their hunting is not only cooperative and strategic; it is also very effective.

In the 1990s the primatologist Craig Stanford set out to study red colobus monkeys, but because so many of them died at the hands of chimpanzees, he ended up switching his study to just that: how and why chimpanzees hunt these red colobus monkeys. He found that chimpanzees were so successful in the hunting of red colobus that the entire social structure of these monkeys was swayed by the annual patterns of chimpanzee hunting. He calculated that some of the most successful communities can bring down nearly a ton of monkey meat in a single year. Subsequent work among some groups of chimpanzees living in west Africa has shown that they even employ tools for hunting, using a specially modified branch spear to kill prey that nest within the holes of tree trunks.

And hunting is by no means restricted to chimpanzees. Related studies among bonobos have been hampered by ongoing (human) wars and the lack of infrastructure in the Democratic Republic of Congo (DRC), the only country in the world with wild bonobo populations. Nevertheless, recent studies have begun to detail the lives of these important relations. Evidence from research conducted over the last ten years or so shows that bonobos, like their chimpanzee (and human) cousins, actively hunt. Some bonobo sites show meat consumption at levels similar to those that have been documented among chimpanzees.

In contrast to humans, chimpanzees, and bonobos, studies of our more distant ape relatives—the gorillas, orangutans, and gibbons—have shown strikingly limited evidence of meat consumption and no evidence to date of hunting. It appears that some of these apes may occasionally scavenge, but even that seems to be quite limited. Taken together the evidence shows that hunting emerged sometime before the split between humans and the lineage that would include chimpanzees and bonobos. Our early common ancestor, living around eight million years ago, probably hunted whatever it could get its hands on but almost certainly hunted the monkeys in the forest habitats in which it lived.

The advent of hunting in these early ancestors surely had

many advantages. The increased caloric intake from hunted animals must have played well in a primarily fruit- and leaf-eating species. The regular supply of monkeys must have increased food stability in a constantly fluctuating food environment. It would have also opened the door for future migration to regions with different kinds of food, a topic to which we will return in chapter 3. Hunting, while undoubtedly beneficial for the first of our ancestors who engaged in it, presents certain undeniable risks for acquiring new and potentially deadly microbes—risks that would continue to have an impact on their descendants for millions of years to come.

Hunting, with all of its messy, bloody activity provides everything infectious agents require to move from one species to another. The minor skirmishes that our early ancestors likely had with other species probably resulted in minor cuts, scratches, and bites—insignificant compared to the intense exposure of one species to another that is a direct result of hunting and butchering.

The chimpanzees who were devouring their feast of red colobus monkey in Kibale forest that day were an instant, visual example of the blurring of lines between species. The manner in which they were ingesting and spreading fresh blood and organs was creating the ideal environment for any infectious agents present in the monkeys to spread to the chimpanzees. The blood, saliva, and feces were spattering into the orifices of their bodies (eyes, noses, mouths, as well as any open sores or cuts on their bodies)—providing the perfect opportunity for direct entry of a virus into their bodies. And since they hunted a range of animals, their exposure to new microbes would have been broad. Those conditions emerged in our ancestors around eight million years ago, forever changing the way that we would interact with the microbes in our world.

While we still only understand the basics of how microbes move through ecosystems, extensive research on toxins gives us an idea of how it works. Microbes, like toxins, have the potential to negotiate their way up through different levels of a food web, a process referred to as *biological magnification*.

Many pregnant woman are aware that there are risks associated with consuming certain kinds of fish during pregnancy. This health suggestion follows from knowledge of how certain chemicals move through food webs. In the complex food webs of the oceans, small crustaceans are consumed by larger fish that are in turn consumed by larger fish and so on. This goes on until we reach the top predator—a hunter who is never hunted—the top of the food chain. Crustaceans have some levels of toxins, such as mercury, that they've accumulated from the environment. The fish that prey on crustaceans accumulate many of these toxins, and the fish that consume these second-order predators accumulate even more. The higher in the food chain we go, the higher the concentrations of such chemicals. So top predator fish like tuna have high enough concentrations of toxins to represent a potential threat to the fetus.

In the same way, animals higher in the food chain should generally be expected to maintain a wider diversity of microbes than those lower on the food chain. They have accumulated microbes like the mercury among fish, in a process we can think of as *microbial magnification*. When our ancestors some eight million years ago took up hunting, they changed the way they would interface with other animals in their environment. And this would mean not only increased interaction with their prey animals. It also meant increased contact with their prey's microbes.

In the twenty years since its discovery, HIV-1 has caused death and illness on a previously unimaginable scale. The AIDS pandemic has affected people in every country in the world. Even today with antiviral drugs that can control HIV, the virus that causes AIDS, it continues to spread, infecting over 33.3 million people at last count. The spread of HIV in contemporary society has a range of determinants, from poverty and access to condoms to cultural practices that dictate whether or not a child is circumcised. The pandemic now has economic and religious significance—and it invites commentary and discussion from philosophers and social activists. Yet it was not always that way.

The history of HIV begins with a relatively simple ecological interaction—the hunting of monkeys by chimpanzees in central Africa. While people normally think about the origins of HIV as occurring sometime during the 1980s, the story actually begins about eight million years ago when our ape ancestors began to hunt.

More precisely, the story of HIV begins with two species of monkey, the red-capped mangabey and the greater spot-nosed guenon of central Africa. They hardly seem the villains at the center of the global AIDS pandemic, yet without them this pandemic would have never occurred. The red-capped mangabey is a small monkey with white cheeks and a shocking splash of red fur on its head. It is a social species living in groups of around ten individuals and eating a diet primarily of fruit. It is listed as vulnerable, meaning its population numbers are threatened. The greater spot-nosed guenon is a tiny monkey, one of the most diminutive of the Old World monkeys. It lives in small groups consisting of one male and multiple females and is able to communicate alarm calls that vary depending on the kind of predator it encounters. One of the things these monkeys share is that they are naturally infected with SIV, the simian immunodeficiency virus. Each monkey has its own particular variant of this virus, something it and its ancestors have probably lived with for millions of years. Another thing these monkeys have in common is that chimpanzees find them very tasty.

L: Lesser white nosed guenon; R: red-capped mangabey. *(L: © Tier und Naturfotografie / SuperStock; R: Shutterstock / Nagel Photography)*

The simian immunodefiency virus is a retrovirus. That means that unlike most forms of life on the planet that use DNA as their code, which translates into RNA and then into the protein building blocks that make up the meat of us all, SIV works in reverse—hence the name "retro" virus. The retrovirus class of viruses begins with RNA genetic code, which is translated into DNA before it can insert itself into the DNA of its host. It then proceeds with its life cycle, creating its viral progeny.

Many African monkeys are infected with SIV, and the red-capped mangabey and greater spot-nosed guenon are among them. While few studies have been conducted on the impact of these viruses on wild monkeys, it is suspected that they do the monkeys no substantial harm. Yet when the viruses move from one host species to the next, they can kill. This would become their destiny.

The work that deciphered the evolutionary history of the chimpanzee SIV was reported in 2003 by my collaborators Beatrice Hahn and Martine Peeters and their colleagues. Over the past decade, Hahn and Peeters have worked tirelessly to chart the evolution of SIV—and they've succeeded. In 2003 they showed that the chimpanzee SIV was in fact a mosaic virus consisting of bits of the red-capped mangabey SIV and bits of the greater spot-nosed guenon SIV. Since SIV has the potential to recombine, or swap, genetic parts, the findings showed that rather than coming from an early chimpanzee ancestor, the virus had jumped into chimpanzees.

It is tempting to imagine a single chimpanzee hunter as *patient zero*—an individual, the first of its species to harbor the novel virus—acquiring these viruses in short order from the monkeys it hunted, possibly on the same day. Alternatively, the mangabey virus may have crossed sometime earlier and gained the ability to spread among chimpanzees sexually, with patient zero acquiring it from another chimpanzee and only subsequently acquiring the guenon virus through hunting. Or perhaps both the guenon and mangabey viruses circulated for some time in chimpanzees after they were acquired through hunting, with the final moment of genetic mixing coming in a chimpanzee already infected by the two viruses. No matter what the particular order of cross-species jumps, at some moment a chimpanzee became infected with both the guenon virus and the mangabey virus. The two viruses recombined, swapping genetic material to create an entirely new mosaic variant—neither mangabey virus nor guenon virus.

This hybrid virus would go on to succeed in a way that neither the mangabey nor guenon virus alone could, spreading throughout the range of chimpanzees and infecting individual chimpanzees from as far west as the Ivory Coast to the sites in East Africa where Jane Goodall began her work in the 1960s. The virus, now known to harm chimpanzees,* would persist

* The same virologist duo—Martine Peeters and Beatrice Hahn—who along with their colleagues showed that SIV was a recombinant of two monkey viruses also showed through long-term monitoring of SIV-infected chimpanzees that, like humans, they also eventually become sick.

in chimpanzee populations for many years before it would jump from chimpanzees to humans some time in the late nineteenth or early twentieth century. And it all started because chimpanzees hunt.

For a large and growing part of humanity, the meat we consume arrives clean and prepackaged, and goes straight to our refrigerators. The killing and butchering of the animals occurs far away on a farm or in a factory that we have never seen and can scarcely imagine. Rarely do we witness blood or body fluids from these animals that were living and breathing beings even a few days earlier. This is because the hunting and butchering of animals is a messy process. We don't want to see it or even think about it; we just want the steak.

During the years I've spent working with people hunting and butchering wild game in places like the DRC and rural Malaysia, I've never become completely accustomed to exactly what is required to prepare meat for consumption. We take for granted what it means to remove hair and skin from a dead animal, the effort needed to separate meat from the many bones distributed in an animal to support its movement. We forget how many parts of an animal must be negotiated to get to the prime cuts: the lungs, the spleen, the cartilage. Watching the process on the dirt floor of a hut or on leaves spread out on the ground in a hunting camp, seeing the blood-covered hands that separate the various parts of the animal and hearing the bits of discarded meat and bone hit the floor still shocks me. It also helps to remind me of the microbial significance of the event.

We tend to think of events like sex or childbirth as intimate, and they certainly bring together individuals in ways that normal interactions cannot. But from the perspective of a microbe, hunting and butchering represent the ultimate intimacy, a connection between one species and all of the various tissues of another, along with the particular microbes that inhabit each one of them.

The butchering in our own kitchen bears little resemblance to the hunting and butchering that our common ancestor would have engaged in eight million years ago. While these first hunting events are now lost, they probably held much in common with the chimpanzees I saw sharing their red colobus meal in Kibale—the dominant male holding down the animal with one hand and using its other hand and teeth to pull apart the skin of the gut while seeking a preferred organ. I remember seeing the chimpanzee holding the organ in its hand, its fur slicked down with blood, and thinking to myself that it would be nearly impossible to imagine a better situation for the movement of a new microbe from one species to the next.

While we still hunt and butcher, the ways that we do so and the methods we use to prepare meat differ radically from the methods of the past. The early ancestors of humans and chimpanzees lacked the ability to cook, they lacked tools for butchering, and they certainly lacked dental hygiene! Whether through a wound from a broken monkey bone, an open sore in the mouth, or a cut on the arm, the microbes of hunted animals infected these animals in ways that had not occurred prior to the advent of hunting. Hunting fundamentally changed how they were exposed to the microbes in their worlds, many of which had remained relatively isolated in the animals that

Chimpanzee eating its hunted prey, a red colobus monkey. (*Nathan Wolfe*)

shared the forests with them. As much as hunting represented a milestone for our eight-million-year-old ancestors, it had equal importance for the world of our microbes.

There are many methods for comparing animals within an ecosystem. We can chart the diversity of foods they consume, the diversity of habitats they utilize, the range of space that they cover within an average year. We can also consider them based on the diversity of microbes they possess, what I call their *microbial repertoire*. Each species has a particular microbial repertoire. It includes viruses, bacteria, parasites—all of the various microbes that can call that species home. And while no single animal within a species will likely have all of the various pieces of the microbial repertoire at any one time, it acts as a conceptual tool for measuring that species' microbial diversity—the range of microbes that infect it.

Species vary considerably in terms of their microbial repertoires. And hunting and butchering do not provide the only avenue for microbes to jump from one species to the next. Species that don't hunt or butcher still have regular exposure to the microbes of other species. Blood-feeding insects provide an important route for microbes to move around. Mosquitoes, for example, often feed on a range of different animals, in effect acting as physical carriers on which microbes can hitch a ride to move from species to species within ecosystems. Similarly, contact with waste from other animals, either through direct contact or indirect contact through water, also provides critical connections in the networks that permit microbes to negotiate the otherwise largely separated worlds of different host species.

Nevertheless, mosquitoes and water provide narrow paths from one host to the next. Mosquitoes, for example, are not syringes. They are fully functional animals that have their own immune systems, and even those microbes that can manage to evade the mosquitoes' defenses will be limited to those in the blood. Similarly, water generally passes on those microbes that live in the digestive tract. Hunting and butchering, in contrast, provide superhighways connecting a hunting species directly with the microbes in every tissue of their prey.

When our ancestors began to hunt and butcher animals, they put themselves at the center of the vast web of microbes living in the full range of tissues of their various prey animals. Whether in the form of a virus in the brain of a bat, a parasite in the liver of a rodent, or a bacterium living on the skin of a primate, the microbial worlds of these various species suddenly converged on the common ancestor, changing for them (and ultimately us) the range of microbes that they would carry.

The impact that the advent of hunting had on the microbial repertoires of the common ancestor and its descendants would continue to play itself out over millions of years. As the lineage of the common ancestor diverged, multiple species (chimpanzees, bonobos, and humans) would emerge, each with the capacity to hunt. These species would go on to accumulate their own sets of novel microbes from the animals on which they preyed. At times, these species would collide when their habitats overlapped, allowing them to exchange microbes, with serious consequences for both species.

Because humans are focused largely on our own health, we often forget that cross-species transmission is not a one-way street. This was brought home for me in vivid detail during my time working with chimpanzees in the Kibale Forest in Uganda. On one afternoon, people from a local village came into our research camp asking for assistance. The distraught villagers explained that a chimpanzee had grabbed an infant child and severely bitten his brother, who had tried to protect it. The infant had not been seen again and was presumably eaten by the chimpanzee. Upon a visit to the village, an eyewitness confirmed the young boy's story. The nasty bite wound on his upper arm was a reminder that would stay with him forever.

The events made me think more carefully about chimpanzee predation, and a subsequent analysis with my colleagues revealed that the event was not unique. Reports from as early as the 1960s had documented similar events. Although it was not a common activity, chimpanzees had hunted humans, usually infants, especially those who were left close to the edge of the forest while their mothers worked on their farms. While disturbing, the idea that chimpanzees occasionally hunt people should not surprise us. From the perspective of a chimpanzee, a red colobus monkey, a forest antelope, and a human infant would all represent logical potential prey. In the same way, humans, while occasionally observing food taboos, hunt opportunistically and generally consume the full variety of local animals in their environment. Whether a closely related ape or a more distantly related antelope, they all present opportunities for vital calories, and both chimpanzees and humans exploit every one of them.

The fact that chimpanzees hunt humans and humans hunt chimpanzees would come to have significance for the two species' microbial repertoires. In the years that followed the advent of hunting by the common ancestor, these two closely related but ecologically distinct species would each accumulate substantive microbial diversity through hunting and other routes. And then, critically, from time to time they would exchange microbes. We'll explore the range of implications that this exchange has in the coming chapters.

As the human lineage broke off and diverged, going through a near extinction event, but then coming back full force with agriculture, animal domestication, and, later, global travel and practices such as blood transfusions, the connections with our ape cousins would continue to have importance for our microbial repertoires in sometimes surprising ways. As we'll explore, the role of this close connection continues now with chimpanzees and other apes acting as the missing piece of the puzzle in some of our most important diseases. Two close primate relatives—chimpanzees that live and hunt diverse animal species in central Africa, and humans with rapidly expanding territory and globally interconnected relationships—would prove to be an important combination. A recipe for pandemics.

➤ 3 ◄

THE GREAT MICROBE BOTTLENECK

We knew it was somewhere nearby, but the area didn't seem quite right. Driving through miles of seemingly endless savanna in Uganda's Queen Elizabeth National Park, we saw only a dozen or so trees, and not the right kind. These were short, solitary, wide-crowned trees, completely engulfed by the never-ending dry grass. Small groups of zebra and the unique Ugandan Kob antelope dotted the landscape. But this did not seem like a a place for a rain forest—certainly not for chimpanzees. It was too open, too dry, too much of a, well, savanna. Yet as we came to the peak of an embankment, there it was—a massive gaping vein of green in the sea of yellow grass. The Kyambura Gorge.

The gorge, while not the only one of its kind, is unusual. Cut through the center by a river flowing from a rain forest some hundred or so miles away, it provides a unique microclimate—a well-hydrated strip in an otherwise dry landscape. Along that strip, rain forest trees and the animals that depend upon them slowly migrated downstream. This occurred over tens and hundreds of thousands of years, but now when you sit in the middle of the usually dry savanna habitat in Queen Elizabeth Park, you see a thriving rain forest, complete with chimpanzees. Effec-

Kyambura Gorge.
(© *Alice Mutasa*)

tively, you see the finger of a faraway forest snaking its way into the savanna.

This gorge provides a unique interface. For contemporary researchers, it provides a fairly easy way to follow chimpanzees. By simply driving along the unobstructed sides of the gorge, we can follow the calls of chimpanzees and then dip down into the gorge to locate them. This is a far cry from the much more challenging work of chasing them through the forest on foot. For the chimpanzees, the gorge provides something much more meaningful. Instead of the few grassland edges in a normal chimpanzee habitat, river gorges like Kyambura provide miles and miles of savanna border along a relatively typical chimpanzee habitat, allowing chimpanzees to explore and utilize the grassland much more than their rain forest contemporaries. And use it they do. Some populations enjoy spending a good deal of time in the savanna, even hunting savanna animals.

Sometime after the split between the lineages that would lead to chimpanzees and bonobos on the one hand and humans on the other, our ancestors embarked upon a trajectory that would include a series of changes moving them away from the lifestyle of the common ancestor. Sitting on the edge of Kyambura Gorge, it is hard not to consider one of the most dramatic of these changes: the shift from being a primarily forest-based animal to an animal with the capacity to live in and utilize grassland. While the order of the events remains somewhat opaque, at some point our ancestors began to make forays into the savanna. This move would ultimately alter their microbial repertoire and their future.

As contemporary humans, we normally think of chimpanzees and bonobos, if we think of them at all, as a side-note species. They are interesting animals, certainly, with much to teach us about our history, yet they hover somewhere near the brink of extinction and live in marginal forest habitats, certainly not species that could compete with humans. Shocking as it may seem, this was not always the case. If we could see the world millions of years ago, say around that time that human lineages and chimpanzee/bonobo lineages diverged, it would be very different. Six million years ago, it was an ape's world.

In our modern world with over six billion humans and only an estimated one to two hundred thousand chimpanzees and ten thousand bonobos, humans have reached every point on Earth, and all of the wild chimpanzees and bonobos remain confined to central Africa. We have to stretch our brains to imagine a world in which we were the minority. Yet for some periods prior to the advent of agriculture around ten thousand years ago, that was exactly the world in which our ancestors lived.

Chimpanzees and bonobos are not fossils. Contemporary

species, whether chimpanzees, bonobos, or humans, have all changed since this aforementioned ancient time. Nevertheless, around six million years ago when our own ancestors took their first tentative steps toward becoming human, they would have seemed much closer to our chimpanzee and bonobo relatives than to ourselves. Our relatives at that time were almost certainly covered with thick hair. When on the ground, they primarily moved around by walking on all fours but really spent the majority of their time in the trees. They hunted—collective and strategic hunting as we've seen. But they didn't cook their meat, didn't use tools that couldn't be simply modified from tree branches, and kept largely to the forest.

As our own lineage changed and began to display some of the features that we equate with humans, the world was a different place. The use of grasslands was perhaps not completely foreign, and even today some small groups of chimpanzees utilize mosaic environments with forest as well as grasslands, like the Kyambura chimpanzees, yet they did not likely make long journeys into these habitats. At that early time, the individuals that spent time in grasslands were the odd ones.

Often when groups of individuals veer into new areas, they do so to escape intense competition, and as our own ancestors moved to savanna habitats, they probably did so less to break new ground than simply to find somewhere they could exist with fewer rivals. Such habitat moves often led to marked inefficiencies, and when our early ancestors relocated, they probably experienced profound disadvantages. Being unsuited, at least at first, to function in grasslands, our early ancestors suffered a number of consequences that likely included smaller population sizes. Or near extinction.

Determining historic population sizes, particularly prior to periods in which we have written records, is fraught with difficulty.

But studies suggest that our ancestors' population densities were sometimes very low, with lower numbers than those of current gorilla and chimpanzee populations, and at least once teetered on the brink of extinction. Our ancestors would have been an endangered species. We believe this to be the case because our genes maintain some of these records, and by comparing the genetic information among contemporary human populations with that of our close ape relatives, we can tease out inklings of relevant information.

The information revealed is striking. Analyses of the human mitochondrial genome, a region of genetic information that passes only from mother to daughter, as well as studies of mobile genetic elements that accumulate in regions of the genome in a clocklike way provide clues to our historic population size and suggest it was much smaller than we might expect.

Our preagricultural ancestors likely lived in small groups, which is not necessarily surprising. Most of our evolutionary history as primates was spent in forested environments. And while exact timelines for the main events remain unknown, moving from forested environments to savanna habitats, shifting from largely fixed territories to a more nomadic lifestyle, and adapting to the various new conditions imposed by these changes must have been traumatic. An apt comparison might be the idea of contemporary humans living on Mars. The generations of our ancestral populations that confronted the savanna frontier probably did so at some cost. But our interest in small population sizes here is less about the consequences for humans and more about the consequences for microbes.

Low population densities, such as those exhibited by our ancestors, have a marked impact on the transmission of infectious agents. Infections need to spread. If population sizes are low, it is much harder for this to happen. The scientific term for substantially reduced population sizes is *population bottlenecks*, and when population bottlenecks occur, species should be expected to lose their microbial diversity.

Microbes can be largely divided into two different groups, acute and chronic, and each group is impaired in small host populations. In the case of acute agents (like measles, poliovirus, and smallpox) infections are brief and lead either to death or immunity from future infections: they kill you or make you stronger. Acute microbes require relatively large populations; otherwise, they will simply burn through the susceptible individuals, leaving only the immune or the dead. In either case, they go extinct. If there's no one left to infect, that's the end of the line for a microbe.

Chronic agents (like HIV and hepatitis C virus), unlike acute agents, do not lead to long-lasting immunity in their hosts. They hang on to their hosts, at times holding on for a host's entire lifetime. These agents have a better capacity than acute agents to survive in small populations. Yet during severe population bottlenecks, even chronic agents suffer from higher rates of extinction. Just like the probability that a particular gene will be lost during a population bottleneck (a phenomenon that results in inbreeding in small populations), the probability that a chronic agent will be lost should also be expected to increase when populations are small. If someone dies, and they are the last individual carrying the microbe, then the microbe dies.

The role of population bottlenecks in diminishing microbial repertoires, what I call *microbial cleansing*, likely had an effect when the population sizes of our ancient ancestors crashed, resulting in populations with a lower diversity of microbial agents. In some cases, microbial cleansing would have led to situations where agents present for millions of years in our ancestors disappeared. Agents that had accumulated following the advent of hunting and other agents that were simply part of our heritage just vanished. While we don't normally think of

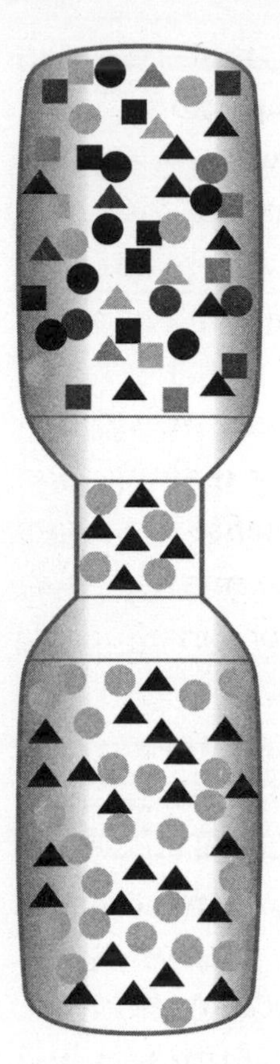

A population bottleneck: a diverse population (top) is greatly diminished by a near-extinction event (middle), resulting in a more homogenous population (bottom). (*Dusty Deyo*)

microbes as a part of our family heritage, in many ways that's exactly what happens—they pass down to us from our ancestors, but from time to time they die out. And while microbial cleansing sounds like a very good thing, it would prove to be a double-edged sword.

Sometime following the split between the chimpanzee/bonobo lineage and our own, another important change occurred in our ancestors that would have dramatic consequences for our microbial repertoires: they learned to cook. Not Michelin three-star cuisine, of course, but cooking nonetheless: using heat to prepare food. Exactly when our ancestors harnessed the power of fire remains a mystery. Presumably, fire first provided warmth and security from predators and competitors. Yet it appears to have quickly become a profound way of altering food. Richard Wrangham, my mentor from Harvard, discusses cooking and its consequences in depth in his well-researched book *Catching Fire: How Cooking Made Us Human*. Among other things, he analyzes in detail cooking's origins.

When our ancestors began to cook extensively, in addition to the advantages that cooking offered them by making food more manageable and palatable, they also benefited from its remarkable ability to kill microbes. While some microbes can survive at incredible temperatures (such as the hot spring microbial hyperthermophiles that grow and reproduce at temperatures above the boiling point of water), the vast majority of microbes that make their living off of animals cannot survive the temperatures associated with cooking. As microbes are heated during cooking, their normally solid, densely packed proteins are made to unfold and open, allowing digestive enzymes quick and easy access to destroy any capacity to function. As with the population bottlenecks that our ancestors swung through, the cooking that became their standard way of life served to again diminish their uptake of new microbes, helping limit their microbial diversity.

The earliest solid evidence that humans controlled fire comes from archaeological finds in northern Israel where burned stone flakes dating back almost eight hundred thousand years were found near fire pits. This is almost certainly an underestimate. African sites dating to over a million years ago contain burned bones that could be the remains of cooking, yet the lack of archaeological evidence makes these finds more ambiguous. In Wrangham's analysis, the evidence of cooking goes back much further. By examining the remains of our ancient ancestors, paleontologists have found physiological clues indicating that they consumed cooked food. For example *Homo erectus*, a human ancestor from 1.8 million years ago, had exactly the larger bodies and smaller digestive tracks and jaws to imply that they consumed higher-energy diets that were easy to chew and easy to digest—in other words, foods that had been cooked.

Whatever the exact date of our ancestors' culinary dawn, it has certainly exploded since then. Cooked foods make up the vast majority of contemporary diets. In my work with hunters around the world, I've had a chance to sample from a vast range of these foods—from roasted porcupine and python in Cameroon to fried wood grubs in rural DRC. On one occasion, my "friendly" Kadazan collaborator in Borneo even gave me dog stew as a practical joke (I didn't really see the humor). I've had a chance to sample food far beyond the beef, lamb, and chicken staples that I grew up eating in America. Yet no matter what I've eaten, or where I've eaten it, one thing can be certain: if the food has been cooked sufficiently, the likelihood that it will make me sick is small.

The dual factors of diminished population sizes and cooking were not the only things that served to decrease the microbial repertoires of our early ancestors. The transition from rain forest habitat to a savanna habitat meant different vegetation and climate but also an entirely different set of animals to interact with and hunt. And different animals meant different microbes.

While we still understand very little about the ecological factors that lead to microbial diversity, there are some key factors that certainly play a role. We know, for example, that the biodiversity of animals, plants, and fungi supported by tropical rain forest systems is higher than any other ecosystem on land. When our ancestors left the rain forest, they entered into regions with diminished biodiversity. The diversity of microbes would almost certainly have been reduced, as would the diversity of the host animals that they infected. So the savanna grassland habitats likely housed fewer animals and a lower diversity of microbes capable of infecting them, which in turn contributed to lower microbial repertoires for our ancestors.

The kinds of animals living in the savanna also differed in critical ways from those in the forests, including a marked contrast in the diversity of apes and other primates. Simply put—primates love forests. The king of the jungle is a primate, not a lion. While some primates, like baboons and vervet monkeys, live very successfully in savanna habitats, forest regions trump savanna regions in terms of primate diversity. When we consider the microbes that could most easily infect our ancestors, the diversity of primates in any given habitat plays an important role. They are certainly not the only species that contribute to our microbial repertoires—in my own studies, I focus not only on primates but also on bats and rodents—but they do play an important role.

Some years ago, I began considering what factors might improve or decrease the chances that a microbe would jump from one host to the next successfully enough to catch on and spread in the new host. It may seem that bats and snakes, for example, would provide similar sources for novel microbes. Yet there is a strong argument against this idea. Long evident to those doing work on microbes in laboratories is the fact that closely related animals have similar susceptibility to certain infectious agents. So a mammal, like a bat, would have many more microbes that could be successfully shared with a human than a snake. If not for the logistics and ethics, chimpanzees would make the ideal models for studying just about every human infectious disease. As our closest living relatives, they have nearly identical susceptibility to the microbes that infect us. Over time, less and less laboratory research on human microbes is conducted in chimpanzees, but this is largely because of the valid ethical concerns associated with conducting research on them and the difficulty of controlling these large and aggressive animals in captivity.

Closely related animal species will share similar immune systems, physiologies, cell types, and behaviors, making them vulnerable to the same groups of infectious agents. In fact, the taxonomic barriers that we place on species are constructs of our own scientific systems, not nature. Viruses don't read field guides. If two different hosts share sufficiently similar bodies and immune systems, the bug will move between them irrespective of how a museum curator would separate them. I named this concept the academically accurate but unwieldy *taxonomic transmission rule*, and it holds up for chimpanzees and humans as it would for dogs and wolves.* The idea is that the more closely related any two species are, the higher the probability that a microbe can successfully jump between them.

Most of the major diseases of humans originated at some point in animals, something I analyzed in a paper for *Nature*, written with colleagues in 2007. We found that among those for which we can easily trace an animal origin, virtually all came from warm-blooded vertebrates, primarily from our own group, the mammals, which includes the primary subjects of my own research, the primates, bats, and rodents. In the case of primates, while they constitute only 0.5 percent of all vertebrate species, they seeded nearly 20 percent of major infectious diseases in humans. When we divided the number of animal species in each of the following groups by the number of major human diseases

* The genetic similarity that dogs and wolves share is virtually identical to the genetic similarity between humans and chimpanzees. For many, this is shocking since we perceive ourselves as so different from chimpanzees yet view dogs and wolves as essentially the same. Such perceptions are more telling of our sensitivity to differences among beings similar to ourselves than they are of the actual genetic relationships between species.

they contributed, we obtained a ratio that expresses the importance of each group for seeding human disease. The numbers are striking: 0.2 for apes, 0.017 for the other nonhuman primates, 0.003 for mammals other than primates, and a number approaching 0 for animals other than vertebrates. So as our early ancestors left the primate-packed rain forests and spent more time with lower overall primate biodiversity in savanna habitats, they moved into regions that likely had a lower diversity of relevant microbes.

Multiple factors likely conspired to decrease the microbial repertoires of our early ancestors. As they spent more time in savanna habitats, our early ancestors interacted with fewer host species, and those hosts were on average more distantly related to them. The advent of cooking increased the safety of meat consumption and stopped many of the microbes that would have normally crossed over during the course of hunting, butchering, and ingesting raw meat. And the population bottlenecks that our ancestors went through further winnowed down the diversity of microbes that already infected them. All in all, the conditions associated with becoming human served to decrease the diversity of microbes present in our ancient relatives. Though many microbes undoubtedly remained in our early ancestors, there were likely far fewer than those that were retained in the separate lineages of our ape relatives.

During the time that our own ancestors went through their microbial cleansing, their ape cousins continued to hunt and accumulate novel microbes. They also maintained microbes that would have been lost in our own lineage. From a human perspective, the ape lineages served as a repository for the agents we'd lose—a microbial Noah's ark of sorts, preserving the bugs that would disappear from our own bloodlines. These great ape* repositories would collide with expanding human populations many centuries later, leading to the emergence of some of our most important human diseases.

Perhaps the single most devastating infectious disease that afflicts humans today is malaria.† Spread by mosquitoes, it is estimated to kill a staggering two million people each year. Malaria has had such a profound impact on humanity that our own genes maintain its legacy in the form of sickle cell disease. Sickle cell, a genetic disease, exists because its carriers are protected from malaria. Protection was so important that natural selection maintained it despite the debilitating disease that appears in approximately 25 percent of the offspring of couples that each carry the gene. People who are afflicted with sickle cell have their origins almost exclusively in one of the world's most intensely malaria affected areas—west central Africa.

My interest in malaria is both personal and professional. During my time working in malaria-infested areas of Southeast Asia and central Africa, I was infected by it on three separate occasions. On the last of those occasions, I almost died. The first two times I'd had malaria were both in regions where malaria was common. I'd exhibited all the typical symptoms—severe

* Sadly, we do not have equivalent information about the microbial repertoires of all of our ape cousins. For example, because bonobos have smaller numbers and live exclusively in the Democratic Republic of Congo, their territory was often inaccessible during the wars of the last twenty years, so we understand much less about their microbial repertoires than we do for those of chimpanzees. As studies of these fascinating apes increase, they will certainly provide additional vital clues to the origins of our own infectious diseases.

† In fact, humans are actually infected with multiple malaria parasites, each with its own evolutionary history. Here, when I refer to *malaria*, I'll use it to mean *Plasmodium falciparum*, the malaria parasite that accounts for the vast majority of human illness.

neck ache (similar to how you'd feel if you slept in a strained position) followed by intense fever and profuse sweating. On each of my first two bouts, I simply went to a local doctor and received a quick diagnosis and treatment. While the pain and illness were miserable, they both resolved reasonably quickly.

I was in complete denial at the time I had my third round with this deadly disease. I wasn't in the tropics; I was in Baltimore! I had returned from Cameroon to do research at Johns Hopkins University, and I had very different symptoms, led by intense abdominal pain. I must have also had fever since I remember complaining to friends who were putting me up in their local bed and breakfast that my room was too cold. These new symptoms and the fact that I'd left Africa many weeks earlier fed my denial that this could possibly be malaria. I finally realized I needed urgent care while sitting half delirious in a tub of scalding water and watching the overflow hit the floor of my friends' bathroom. Although I recovered after a few days in the hospital, the illness brought home for me the huge impact that this disease has on the millions of people who are regularly sickened by it.

My professional interest in malaria had started much earlier. As a doctoral student studying the malaria of orangutans in Borneo, I'd had the good fortune to spend a year working with some of the world's foremost experts on malaria evolution at the CDC in Atlanta. There I had the luxury of spending afternoons with Bill Collins, perhaps the world's greatest expert on the malaria parasites of primates, discussing how malaria might have originated. Among the prominent themes of our chats was the importance of wild apes.

At the time, we knew that wild apes had a number of seemingly distinct malaria parasites. One of them was particularly intriguing. *Plasmodium reichenowi* was named after a famous German parasitologist, Eduard Reichenow, who had first documented the parasites in chimpanzees and gorillas in central

Africa. Reichenow and his contemporaries saw a number of these particular parasites, collector's items for the German researcher, and correctly identified them through examination by microscope as closely related to our own *Plasmodium falciparum*. In the 1990s, during my time at the CDC, molecular techniques were paving the way to detailed examination of these parasites, allowing us to compare them accurately to our own parasites and providing much greater evolutionary resolution than a microscope could ever offer. Sadly, all of the parasites of Reichenow's time had been lost, and all that remained was a single lone specimen.

Initial work with this lone *P. reichenowi* parasite showed that in fact it was the closest of the many primate malarias to our own deadly human malaria, *P. falciparum*. Yet with only a single specimen, it remained impossible to say much about the origins of these parasites. Perhaps, long ago, the common ancestor had a parasite that over millions of years had gradually evolved into distinct lineages of *P. reichenowi* and *P. falciparum*, a hypothesis favored by some at the time. Or perhaps the ape parasite simply resulted from the transmission of the common human parasite to wild apes at some point in fairly recent evolutionary history. A third possibility, neglected by most considering the huge number of humans and the incredible proliferation of *P. falciparum* among them compared to the existence of only a few dozen known parasites in apes, was that perhaps *P. falciparum* was in fact an ape parasite that had moved over to human populations.

Bill and I understood that to truly address the evolutionary history of these parasites we'd need to get more samples from wild apes, ideally many. As a young doctoral student, I was ambitious yet still naïve about the difficulties associated with getting these kinds of samples. But I promised Bill I'd do it and set about planning ways to sample apes in the wild.

Unbeknownst to me at the time, I was about to be called

away by my soon-to-be postdoctoral mentor Don Burke to conduct research in Cameroon. I was unaware at the time that I'd spend nearly five years establishing a long-term infectious-disease-monitoring site there in Cameroon. Eventually, though, I did follow through on my promise to Bill and got those ape samples. Ultimately, in collaboration with sanctuaries in Cameroon that helped to provide homes to orphan chimpanzees, we discovered that ape malaria parasites were not as uncommon as people had suspected. By teaming up with Fabian Leendertz, a veterinary virologist who had done similar work in the Ivory Coast, molecular parasitologist Steve Rich, and the legendary evolutionary biologist Francisco Ayala, we took an important step toward cracking the origin of this disease.

Together we were able to compare the genes in hundreds of human *P. falciparum* samples that already existed with around eight new *P. reichenowi* specimens from chimpanzees in locations throughout west Africa. The genetic comparison surprised us all. Amazingly, we found that the entire diversity of *P. falciparum* (the human malaria) was dwarfed by the diversity of the handful of *P. reichenowi* chimpanzee parasites we'd managed to uncover. This discovery told us that the most compelling explanation for *P. falciparum* was that it had been an ape parasite and only jumped over to humans through a bite by some confused mosquito, sometime after our split with the chimpanzee lineage. Human malaria had, in fact, originated in wild apes. In the years that followed our work, a number of researchers documented more and more of the parasites in wild apes.

Subsequent work by my collaborators Beatrice Hahn and Martine Peeters (the same scientists who have done work on SIV evolution) has shown that the malaria parasites infecting wild apes are even more diverse than our study indicated. They have shown that the ape parasites most closely related to human *P. falciparum* exist in wild gorillas, rather than chimpanzees. How these parasites have been maintained among wild apes and whether or not they've moved back and forth between chimpanzees and gorillas remain questions for future studies. Either way, there is no longer any doubt that human *P. falciparum* moved from wild apes into humans and not in the opposite direction.

That malaria crossed from a wild ape into humans makes great sense when viewed from the perspective of the evolution of our lineage. The microbial cleansing that resulted from habitat change, cooking, and population bottlenecks among our own ancestors had cleared our microbial slate, decreasing the diversity of microbes that were present before. Perhaps the many years with leaner microbial repertories had also decreased selective pressure on the many innate mechanisms that we have to fight against infectious diseases, effectively robbing us of some of our protective disease-fighting tactics.

In more recent times, as our population sizes began to increase, wild ape diseases, some of which we'd lost millions of years earlier, had the potential to infect us again. When these diseases reentered humans, they acted on us like uniquely suited novel agents. Malaria was not the sole microbe to make the leap from apes to modern humans, and the stories of others, like HIV, tell a strikingly similar tale. The loss of microbial diversity in our early ancestors and the resulting decrease in their genetic defenses would make us susceptible to the microbial repositories that our ape cousins maintained during our own microbial cleansing. While we continued to change as a species, yet another part of the stage would be set for the brewing viral storm.

▸ 4 ◂

CHURN, CHURN, CHURN

The oysters were excellent, but the company was even more striking. As I sat in the small Parisian bistro with a tray of fresh shellfish, I savored the taste of the ocean. But the more powerful memory of that day was of another patron of the restaurant. At the table next to me sat an impeccably put together Frenchwoman. Her bag, skirt, and socks all matched—not exactly, but just enough to notice. Her dining companion sat to her right—a miniature poodle, sitting on the chair and drinking water from a bowl on the table. Pieces of his meal—chicken I think—fell over the side of his plate, mingling with the crumbs from his owner's bread.

Dogs play an important role in the lives of many people around the world. I had stopped only briefly in Paris on the way home from a month-long trip conducting research in Asia and Africa. It might have been the jet lag, but my recollection of the event could only be described as surreal. During my trip I'd spent time in a part of Borneo where people eat dog, including on at least one occasion my unsuspecting self. I'd also visited Muslim areas of the Malay Peninsula, where devout people won't even touch dogs because of religious beliefs. And I'd spent time in central Africa, where I'd seen local hunters work with their small, silent basenji hunting dogs—dogs that lived on their own but in exchange for scraps followed hunters into the forests, helping them catch their prey. In the United States, many people treat dogs as members of their families, paying large fees for medical expenses and mourning for them when they die. Sitting on the beach near my home in San Francisco, it would be hard for me to spend an hour without seeing someone kiss his or her pet dog on the mouth. Watching that woman in Paris sharing a meal with her dog solidified just how linked we are to these animals.

The close relationships we have with dogs, whether as companions, work animals, dinner guests, or a source of food, should not surprise us. Dogs play a special role in human history. If we were to compile the "greatest hits" of human evolution, hunting and cooking would certainly make the cut. Language and the capacity to walk on two feet would also be on the list. But central among our species' critical historical events is domestication—and dogs were the first in a long line of plants and animals that our ancestors tamed.

The capacity to domesticate plants and animals underlies much of what we now think of as being human. To imagine a world without domestication, we'd have to spend time with one of the few dozen human populations on the planet that still practice hunting and gathering lifestyles, groups like the Baka and Bakoli, the so-called pygmies, living in central Africa that I have worked with for years, or the Aché that live in South America. For these groups of people, there is no bread, no rice, no cheese. There is no agriculture, and therefore the many rituals of our planet's major traditions, including the harvest and planting pilgrimages and their associated festivals, are entirely absent—no holidays such as Ramadan, Easter, or Thanksgiving.

There is no wool, no cotton, only textiles made from wild tree bark or grasses and the skins from hunted animals.

These hunter-gatherer populations have complex histories, and many of them lived at some point with some form of agriculture before returning to a foraging lifestyle. Yet they provide us with interesting clues on what the lives of our ancestors looked like before the advent of widespread domestication.* Among the traits hunter-gatherer populations share are small population sizes and a nomadic lifestyle. As we'll see, these traits have an important impact on keeping the microbial repertoires of these populations at low levels.

The first human foray into domestication came with modification of wolves into the canines we know today. Archaeological and DNA evidence suggests that populations in the Middle East and east Asia began domesticating gray wolves as early as thirty thousand years ago, turning them into guard dogs and work animals as well as using them for food and fur. The early history of dog domestication is still unclear. One hypothesis is that wolves followed humans, scavenging off of their kills, and over time became dependent on humans, a dependency that set the stage for their later domestication. But no matter how it began, by fourteen thousand years ago dogs played an integral role in human life and culture. In some archaeological sites in Israel, humans and dogs were even buried together. These early dogs would have resembled modern-day basenjis, the silent hunting dogs preferred by the central African hunters with whom I work.

Occurring around twelve thousand years before we would

* Unlike our ancestors forty thousand years ago who lived with no animals for protection or to assist with labor, all current hunter-gatherer populations have dogs.

Female Basenji dog. (*Dave King / Getty Images*)

domesticate anything else, the domestication of the dog was an early precursor to what would follow. Around ten to twelve thousand years ago, a *domestication revolution* occurred in earnest, starting with sheep and rye and then followed by a diverse group of other plants and animals.

The consequences and opportunities of the domestication revolution were profound. Prior to domestication, human populations were limited by the food available in wild environments. Wild animals migrate, which forced our ancestors, who were dependent on the hunting of these wild animals, to do the same. The wild fruits and other plant foods present in the local habitat were spread out, which again forced seasonal movement. Wild environments, with a few minor exceptions,*

---

* One notable exception was populations supported by marine habitats. People living off of the ocean through fishing and the hunting of marine mammals were often able to achieve relatively large population sizes and maintain a sedentary lifestyle without domestication. While likely not sustainable in the long-term, the vast quantities of animal protein present in certain marine systems mimicked the concentrated caloric resources that subsequent domestication would provide.

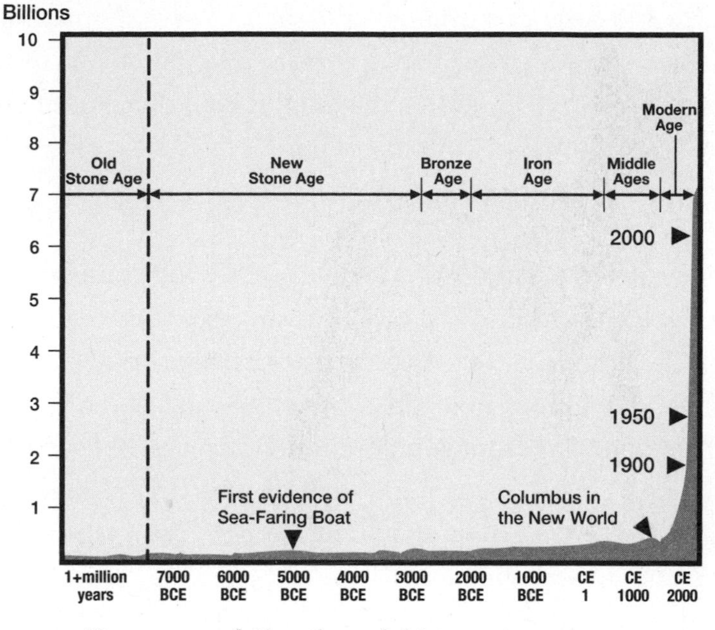

Human population through history. (*Dusty Deyo*)

lacked the capacity to sustain large populations of people. As a consequence, human population sizes were small, probably numbering no more than fifty to a hundred people in a group, and mobile.

As domestication truly kicked in around five to ten thousand years ago, this would all change. With a combination of domesticated plants and animals, humans gained the capacity to have sustained sources of calories year-round. Agriculture (i.e., the domestication of plants) made it possible for human populations to stay in one place and avoid the constant movement that characterizes hunting-and-gathering populations as well as populations with only domesticated animals, which need to move in order to find feed for their herds. A sedentary lifestyle and the capacity for food surplus radically increased the potential for populations to grow, leading to the first real towns and cities. The particular combination of larger population sizes, sedentary groups of humans, and the growing populations of domestic animals would play a central role in transforming the relationship between humans and microbes. But humans aren't the only animals that tame the wild.

Despite conventional wisdom, the capacity for domestication is not unique to humans. The most striking example of domestication in the animal kingdom comes not from primates, dolphins, or elephants—in fact, not from a vertebrate species of any type—but from ants. Far from simple-minded insects, ants are part of unique and complex colonies, each of which is perhaps better imagined not as a group of individual ants but rather as a collective ant "superorganism."*

Leaf-cutter ant colonies exist in most tropical American habitats. Known to schoolchildren worldwide for their incredible strength, the workers march through the jungle carrying pieces of green leaves many times their own size back to the nest. Yet the leaf-cutter's strength is not its most interesting feature. This amazing group of ants has mastered the art of domestication. Rather than eat those massive leaves, the workers chew them up into a fertilizer. The colony uses the fertilizer in order to support their gardens—for leaf-cutter ants, made up of the *Atta* and

---

* Ant societies, like bees, consist of large colonies of female workers, all descended from a single mother (the queen) and father. The workers' fathers (and all males) result from the development of unfertilized eggs, which means they lack half of the genetic information of fertilized offspring, or are haploid, in scientific terms. The haploid father contributes identical genetic information to each of his daughters. For this reason, the worker ants in a colony share 75 percent of their genetic information, rather than the 50 percent shared by sisters in species like our own. Because of their close genetic relationship with each other, female workers lie squarely in the middle of the continuum between sisters and cells. Ants in a colony are more accurately thought of as physically distinct cells in a large and single organism (i.e., the colony/hive) rather than as collectives of cooperating unrelated individuals.

Leafcutter ants in a fungus garden, Belize. (*Mark Moffett / Getty Images*)

Acromyrmex groups, cultivate a fungus-based crop and have spent millions of years living off it. These ants are farmers.

Domestication of fungus has helped leaf-cutter ants become one of the most successful species on our planet. Mature leaf-cutter colonies, measuring fifteen meters across and five meters deep, can house upward of eight million ants. The massive underground colonies are sedentary, sometimes lasting for more than twenty years in the same location.

These remarkable ants have attracted a number of scientists, including a Canadian researcher named Cameron Currie. Dr. Currie has used molecular tools to examine the genetics of the ants, their fungus, and the other members of this incredible community. His research has shown the evolutionary links between the ants and their fungus crop. The colonies and their crop species have lived together for tens of millions of years, a much more mature farmer-crop relationship than that seen in humans.

Like human farmers, the ants have agricultural pests, including a specialized fungal parasite that spoils the farms.

Dr. Currie has shown that not only have the ants and their crops lived together for millennia; the parasitic fungus has been along for the ride since the beginning. Another amazing twist to this elegant system is that, like human farmers, the ants utilize a pesticide. They cultivate a species of bacteria that produces fungicidal chemicals that help the ants control their vermin. Some people think of ants as pests, but these ants have their own pest problems.

Humans began domesticating other species merely thousands of years ago, rather than millions, as with the leaf-cutters. Like the ants, we've found that one of the consequences of high crop densities is parasites. The fungus species that the ants cultivate almost certainly had pests tens of millions of years ago, before they were cultivated by the leaf-cutters. But when the leaf-cutters accumulated the fungus and added fertilizer, it allowed more fungus to live closer together than resources would have permitted without active farming. Cultivation leads to concentrated populations, and concentrated populations have higher burdens of parasites, whether fungus or virus.

While the leaf-cutters focus exclusively on farming fungus, humans have taken agriculture and livestock to entirely new levels. Rather than cultivate a species or two over the course of a few millennia—lightning speed in evolutionary terms—humans domesticated a vast range of plant and animal species.

We take it for granted, but the diversity of living things that our species cultivates boggles the mind. In an average day, we might wake up in sheets (cotton) and wool blankets (sheep); put on leather shoes (cow) and perhaps a cashmere sweater (goat); eat a breakfast of eggs (chicken) and bacon (pig); bid farewell to our pets (dog, cat) on the way to work; for lunch we might eat a salad (lettuce, celery, beets, cucumber, garbanzo beans, sunflower seeds) with dressing (oil from olives); for a snack we might eat a fruit salad (pineapple, peaches, cherries, passionfruit) or mixed nuts (cashews, almonds, peanuts, actually a legume); for dinner a caprese salad (tomato, buffalo mozzarella) and pasta (wheat) with peas and smoked farmed salmon with fresh basil (all domesticated). It would be an uncommon day for many of us *not* to interact with at least three domesticated animals and a dozen or so domesticated plants. We are truly masters of domestication.

Consumption of wild foods, the source of calories for virtually all other organisms on our planet, now represents an almost quaint luxury for most humans. My friends Noele and Giovanni make a delicious wild asparagus pate from plants gathered in woods outside their small hillside village near Reggio, Italy. But using wild vegetables is now the exception rather than the rule. Wild salmon costs significantly more than farmed salmon in the vast majority of the world. Eating wild venison, something my friends Mimi and Chris like to do each year in their Massachusetts cabin, represents a challenging "return to nature" rather than a regular source of calories.

The transition from a species primarily dependent on wild sources of nutrients to a species that cultivates most of its food means that we don't need to depend on the fluctuating food availability in uncultivated habitats. It also allows for the concentration of these activities, with a few individuals focused on developing food while the rest of us have time to pursue other objectives, like, say, virology. We are freed from the daily foraging required of our ancestors before domestication. For our purposes here, it also radically changed the way that we related to the microbes in our world.

In the field sites where I work throughout the world, my collaborators and I work closely with hunters and monitor for new microbes that cross into them as they catch, prepare, and consume wild animals. Yet the hunters are not our only focus. Among the things we study in rural villages are the domestic animals—the dogs, goats, pigs, and other species that surround these people. Each animal, wild and domestic, has their own microbial repertoire, and when concentrated on a farm or in a house or herd, these microbes thrive.

Domestic animals have contributed novel microbes to humans in different ways. Since these species each had their own pre-domestication microbial repertoires, the initial close contact of farming led to an early exchange of their microbes to humans. My colleague Jared Diamond has provided detailed evidence for this exchange and its consequences for human history in his excellent book *Guns, Germs, and Steel.* Among other things, Jared showed that the preponderance of domestic animals in temperate regions contributed to a higher diversity of microbes among temperate populations. For example, measles descends from rinderpest, a virus of cows that entered into humans, a domestication-associated virus that continues to plague us.

Humans have close interactions with domesticated animals, whether for companionship, protection, or food. These interactions reach fascinating extremes. In Papua New Guinea, women in some ethnic groups actually suckle their pigs, providing human breast milk to ensure the survival of these valuable animals. This level of close connection has obvious implications for the movement of infectious agents.

Of the microbes that originated in our domesticated animals, many entered into humans thousands of years ago, at or near the time that we first domesticated them. Acquiring the microbes that belonged to our domestic animals played an important role in enhancing the microbial repertoire of our ancestors during the climax of domestication five to ten thousand years ago. Over time, this has changed. In the case of dogs, for example, most of the microbes that they had to contribute to humanity have already crossed over. In some ways, the microbial repertoire of our species has merged with that of dogs and the other animals we've domesticated. Even without breastfeeding our domestic animals, we often cuddle with them for warmth or play. We almost always have closer connections to them than we would to wild animals.

The historical "predomestication" dog microbes that had the potential to cross into humans have largely done so, and the human microbes that could survive in dogs have also crossed. The ones that haven't crossed successfully likely don't have the potential to, and while they may lead to occasional infections in one or two individuals, they won't have the capacity to spread—the critical trait required for something to have true impact.

Over the thousands of years of interaction, we have reached a sort of microbial equilibrium with domestic animals. But this doesn't mean that these animals don't still contribute to our microbial repertoire; quite the contrary. Domestic animals continue to feed new microbes into the human species. These bugs derive not from the animals themselves, but from wild animal species that they are exposed to. Our domestic animals act as microbial bridges, permitting new agents from wild animals to make the jump into us.

There are numerous examples of domestic animals bridging the microbial divide between humans and wild animals. Perhaps the best documented of these is the case of Nipah virus, a fascinating bug whose emergence has been studied in detail by my collaborators Peter Daszak and Hume Field and their colleagues. Through years of viral sleuthing, they have shown in exquisite detail exactly how the virus negotiates the complex world of humans and our farms.

Nipah virus was first detected in Malaysia, in the village that gave it its name. This virus kills. Of the 257 cases of infection seen during 1999 in Malaysia and Singapore, 100 people died, a startlingly high mortality rate. Among the survivors, more than 50 percent were left with serious brain damage.

The first clues to the origin of the virus were the patterns of human cases. The vast majority occurred among workers in piggeries. At first, the investigators thought the virus causing the illness was Japanese encephalitis virus, a mosquito-borne virus present throughout tropical Asia. Yet menacing and distinct symptoms led the investigating teams to determine that it must be a new and still unidentified agent.

Early symptoms of Nipah virus include those common in viral infections—fever, decreased appetite, vomiting, and flu-like systems. But after three to four days, more serious nervous system manifestations appear. The exact impact that the virus has differs from person to person. Some individuals experience paralysis and coma, while others have hallucinations. One of the first documented patients reported seeing pigs running around his hospital bed.

MRI scans show serious damage to patches of the brain, and the patients who die usually do so within a few days of the onset of brain damage. Among the individuals infected in Malaysia and Singapore in 1999, none appeared to seed additional human infections, yet cases in subsequent years in Bangladesh provide evidence that the virus has the potential to spread from human to human under at least some circumstances.

When scientists discover a new virus, a mad rush often ensues to identify the *reservoir* of the virus—the animal that maintains it. While certainly useful, the concept of a reservoir also has limitations. Scientists often see stark divisions between species. We neatly divide up the world of animals into families, genera, and species, but we often forget that these divisions are based on our own conventions. A taxonomist can clearly sort out the difference between a colobus monkey, a baboon, a chimpanzee, a gorilla, and a human, yet the traits that permit us to classify these animals as distinct are, as I've mentioned, often irrelevant for a microbe. From the perspective of a virus, if cells from distinct species share the appropriate receptors, and ecological connections provide the appropriate opportunities to make a jump, the fur of a baboon or the upright status of a human does not matter at all.

Some viruses persist permanently and simultaneously in multiple hosts. Dengue virus, a viral infection originally called breakbone fever because of the intense pain it causes, appears largely in human cities. Yet dengue also lives in wild primates in tropical forests, where it is referred to as sylvatic dengue.* Sylvatic dengue simultaneously infects multiple species of primates and does not discriminate. It has a wide *host range*.

Among the numerous dry technical scientific papers that I digested as a doctoral student, few are indelibly etched on my brain. One that I remember in detail was a report describing experiments to determine the host range of sylvatic dengue.

In the study, which used outdated methods now considered unethical, scientists put various species of primate into cages and used ropes to lift the cages high into the canopy where dengue's forest mosquitoes feed. There they gathered samples of viruses to determine which species had the potential for infection. The study largely worked—except in one case where they brought the cage down only to find a massive python with a very badly distended abdomen. The large snake had entered

* There is ongoing debate among scientists as to the importance of sylvatic dengue for the human outbreaks. Unfortunately, the intense difficulties associated with isolation of dengue virus from forest settings makes ideal comparisons a challenge.

Image from the sylvatic dengue study. (*Institute of Medical Research Malaysia / A. Rudnick, T. Lim*)

the cage to consume the trapped and no doubt terrified monkey. Having miscalculated, the satiated snake could not squeeze through the bars to escape and found itself in the same trapped predicament as its monkey prey. Most likely the snake didn't get infected with the virus; few viruses infect both reptiles and mammals. It did, however, make for a memorable photo in an otherwise dry technical journal.

The capacity for sylvatic dengue to thrive in multiple species presumably helps the virus persist in regions where the population density of any single primate species would not be sufficient to protect the virus from extinction. And the mechanism dengue uses to move from one animal to another—mosquitoes—helps make this movement seamless.

For dengue, the notion of a single reservoir does not, strictly speaking, make sense, but when Nipah was discovered in 1999, that was still unclear. Scientists then asked themselves: what local animal or animals, wild or domestic, were Nipah's reservoir? Knowing what animal or animals a virus lives in prior to infecting humans helps us respond to it. Depending on the reservoir, we may have the potential to simply change farming practices or modify human behavior to avoid the critical contact that leads to viral exchanges, effectively cutting off the virus's ability to enter humans.

Knowing that a microbe has the capacity to maintain itself in an animal reservoir also changes the way that we think about public health strategies. Microbes can jump in both directions, so while novel human microbes like Nipah originate in animals, established microbes also have the potential to cross back into animals. Animal reservoirs for established human bugs can potentially derail control efforts. In effect, if we eliminate a bug in humans in a particular region, but it lives on in animals, the microbe may have the potential to reemerge with deadly consequences. In order to truly eradicate a human pathogen, we must know if it can also live outside of humans.

When Nipah emerged in 1999, the scientists studying it moved quickly to home in on its reservoir. Over the years that followed, an intricate relationship among wild animals, domesticated animals, and plants revealed itself, a story that emphasizes the complex ways that domestication can provide new avenues for bugs to pass into people.

The Malaysian piggeries that Nipah entered are not small-scale affairs. They house thousands of pigs at very high densities, creating a ripe environment for viral spread. The farmers who raise the pigs work hard to maximize their income both from the pigs themselves, but also from the surrounding land. One of the practices in this area of southern Malaysia is to grow mango trees in and around piggeries, providing a second source of income to increase the viability of the farming enterprise.

In addition to producing delicious fruit for the farmers to sell, the mango trees attract the flying fox, a large and appropriately

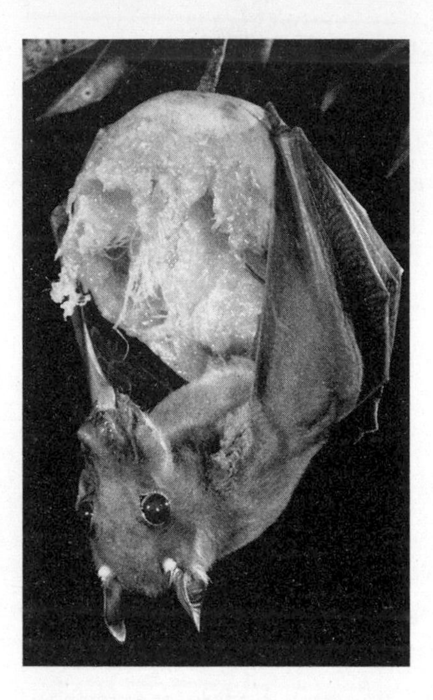

Wahlberg's Epauletted Fruit Bat (*Epomophorus wahlbergi*) eating mango. (*Dr. Merlin D. Tuttle / Bat Conservation International / Photo Researchers, Inc.*)

named bat with the scientific name *Pteropus*. This bat was the unexpected Nipah reservoir, the virus's link to the wild. Remarkably, it now appears that the *Pteropus* bats, while consuming their mango suppers, urinate and drop partially eaten mango into the pig pens. The omnivorous pigs consume the Nipah-infected bat saliva and urine as they eat the mango. The virus then spreads quickly in the dense pig populations, which, because the animals are sometimes shipped from place to place, infect new piggeries and occasionally infect their human handlers.*

Emerging thousands of years after the advent of domestication, Nipah illustrates the impact that domestication had on

* Recent work in Bangladesh by researchers at the International Center for Diarrheal Disease Research has shown that Nipah can enter into humans without pigs. One of the delicacies in parts of the country is sap from date palm trees, which is tapped overnight and consumed fresh in the morning. During the night, bats feed on the sap that flows into collecting pots, on occasion contaminating the sap with Nipah virus.

our relationship with microbes. The larger and more sedentary populations of humans that emerged following the domestication revolution were susceptible to outbreaks in ways that our predomestic ancestors never were. In the small mobile communities that dominated human life prior to agriculture, novel microbes that entered these communities from animals would often sweep through, killing certain individuals and leaving the rest of the small populations immune. At that point the viruses would effectively die out; a virus without a susceptible host is unable to survive.

As villages and towns formed around agricultural centers, they did not do so in isolation. Communities were connected, at first with footpaths, then roads. While we might think that these towns were separate functional entities, from the perspective of a microbe, they represented a single larger community. As this interconnected community of towns grew, it provided the first opportunity in human history for an acute virus to persist permanently in the human species.

Chronic viruses that live permanently within their hosts, like hepatitis B, do not necessarily require large populations because they can continue to pass on their progeny for many years. These viruses have the potential to persist in very small communities, taking a long-term strategy—he who fights and hides away lives to fight another day. On the other hand, acute viruses, such as measles, do not remain in a single individual for long and require a constant supply of susceptible hosts. As they burn through populations, they kill some and make the rest immune, often leaving no one to perpetuate the infection.

Therefore, within the small, mobile hunter-gatherer lifestyle that our ancestors led prior to domestication, acute viruses could not survive for long unless they were microbes that we shared with other species. In the same way, chimpanzee populations, including those that were studied by the pioneering primatologist Jane Goodall, have sometimes been hit with polio. The virus that causes polio normally requires large populations of contemporary humans to sustain itself. Nevertheless, in 1966 Dr. Goodall and her colleagues saw that the wild chimpanzees they studied had come down with something that looked very much like human polio, including symptoms of flaccid paralysis. The outbreak was devastating for the chimpanzee community in Tanzania, killing a number of animals.

The virus that caused chimpanzee polio was in fact the same virus that caused polio in humans. It had jumped over *from* nearby humans who were experiencing an outbreak at the same time. Dr. Goodall and her colleagues administered vaccine to the chimpanzees, which no doubt limited the harm to the community. Chimpanzees, like our early human ancestors prior to domestication, would not have had the population sizes to maintain such a virus—current estimates suggest that communities of over 250,000 people are necessary to sustain it. In small communities, the virus would simply have swept through, harming some and creating immunity in the others, before dying out.

But when our ancestors, with their farms and domestic animals, began to have interconnected towns, viruses like polio gained the ability both to infect us and to be maintained within our species. As more and more towns appeared and the connections between them improved, the number of people in contact with each other increased. From the perspective of a microbe, the physical separation of these towns didn't matter if there were enough people moving between the towns. Hundreds, and later thousands, of interconnected towns effectively became a single megatown for microbes. Eventually, the number of interconnected people would become so large that viruses could maintain themselves permanently. As long as new people entered into the populations through birth or migration, and did so with enough frequency, there would always be a new person for the microbe to try.

In effect, domestication provided a triple hit to our ancestors when it came to microbes. It provided sufficiently close contact with a small set of domesticated animals, allowing their microbes to cross over into us. At the same time, domestic animals provided a regular and reliable bridge to wild animals, giving their microbes increased opportunities to cross into us. Finally, and perhaps most crucially, it permitted us to have large and sedentary communities that could sustain microbes that previously would have been a flash in the pan. Together, this viral hat trick put us in a new microbial world—one that would lead, as we'll see in the next chapter, to the first pandemic.

PART II

THE TEMPEST

➤ 5 ➤

THE FIRST PANDEMIC

In early July 2002 in Franklin County, Tennessee, a thirteen-year-old boy named Jeremy Watkins picked up a sickly bat on his way home from a day of fishing. None of the other family members handled it, and his stepfather wisely made him release the animal soon after Jeremy revealed his find.

Events like this happen all over the world with thousands of wild animals every day, largely without ill consequences. But Jeremy's encounter with this particular bat would be quite different.

In the CDC report that would document Jeremy's case, the next events were described with clinical efficiency. On August 21 Jeremy complained of headache and neck pain. Then a day or so later his right arm became numb and he developed a slight fever. Perhaps of greater concern, he also developed diplopia, or double vision, and a constant, queasy confusion. Three days later he was taken to the local hospital's emergency room but was discharged with the incorrect diagnosis of "muscle strain." The next day he was back in the emergency room, this time with a fever of 102°F. He had the same symptoms, but now his speech was slurred, he had a stiff neck, and difficulty swallowing.

At this point, Jeremy was transferred to a local children's hospital. By August 26 he could no longer breathe or think normally. He was also producing copious amounts of saliva. Highly agitated to the point of being combative, Jeremy was sedated and put on life support. His mental status deteriorated rapidly and by the next morning he was completely unresponsive. On August 31 Jeremy was pronounced brain-dead and, following the withdrawal of life support, he died of bat-borne rabies.

Jeremy's family did not know that bats could carry rabies, much less transmit it to humans. They did not remember him complaining of a bite, although that's exactly what must have happened while he carried the bat home from his fishing excursion. They probably did not know that the incubation period for rabies is generally three to seven weeks, well within the range of the time between his exposure to the bat and the first symptoms he experienced. Detailed studies of the virus that killed Jeremy revealed evidence of a variety of rabies found in silver-haired and eastern pipistrelle bats common in Tennessee.

Rabies is a terrible way to die. It's a disease that devastates the families of its victims, with patients becoming virtual zombies in the days before death. It is among the small number of viruses that kill virtually all of the individuals they infect. But as tragic as it is that the doctors at the local Franklin County emergency room sent Jeremy home with a diagnosis of muscle strain, the reality is that it was already too late to help the boy at that point. Without rapid postexposure prophylaxis after infection, the boy was destined to die.

If we take a different view, the virus that causes rabies is not only a deadly menace but also a truly amazing feat of nature. This virus, shaped like a bullet, is a meager 180 nanometers long and 75 nanometers across. If you stacked rabies viruses one on top of the next, you would need more than a thousand of them to reach the thickness of a single human hair. Rabies has an almost trivial genome, with only twelve thousand bits of genetic information for a meager five proteins. It's simple, tiny, and incredibly powerful.

While diminutive, the virus accomplishes remarkably sophisticated tasks. In addition to the standard viral work of invading cells, releasing genes, making new viruses, and spreading, it has some unique tricks. From the point of entry, the virus travels preferentially along neural pathways, making its way into the central nervous system. It accumulates selectively in the saliva. The virus particles that infect the central nervous system modify the host's behavior, increasing aggression, interfering with swallowing, and creating a profound fear of water. When put together, a rabies infection leads to an aggressive host literally foaming at the mouth with virus. A host that lacks the capacity

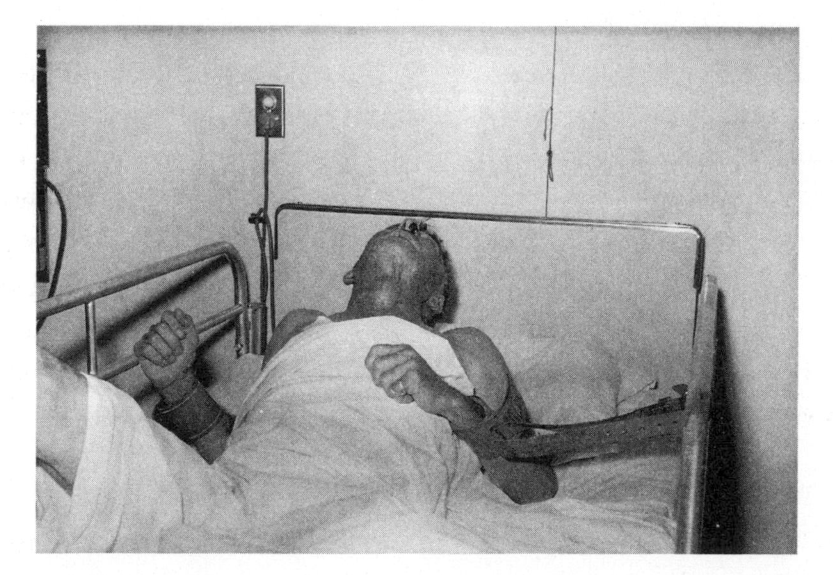

Hospitalized human rabies victim in restraints *(Courtesy of the Centers for Disease Control)*

to drink or swallow further increases the probability of delivering a successful bite—a bite that gives this particular virus the ability to advance from one individual to the next.

As frightening and deadly as rabies is, as a global community we need not fear it. That a virus is exceedingly and dramatically deadly does not mean it will become a pandemic. Rabies kills more than fifty-five thousand people a year worldwide. It is a cause for serious public health measures, but it does not present a global pandemic threat. In all of the years that the CDC and other public health organizations have followed rabies, it has never once gone from person to person. Every one of those deaths, like the death of Jeremy Watkins, resulted from an independent animal infection. From a pandemic perspective, it doesn't have the right stuff.

So, what *is* a pandemic? Defining them creates some trouble. The word itself comes from the Greek *pan*, meaning "all," and *demos*, meaning "people." Yet, in reality, it is almost impossible to imagine an infectious agent that infects the entire human population, a high bar to set for a virus. In humans or any hosts, different individuals will have different genetic susceptibility, so at least a few individuals will likely be incapable of sustaining an infection because of some kind of genetic immunity. Also, the simple logistics of spreading to every single individual in any population makes such a feat nearly impossible.

Among the most common viruses infecting humans that we are aware of is the human papilloma virus,* and it doesn't afflict 100 percent of people. HPV currently infects 30 percent

---

* There is a class of viruses, the endogenous viruses, which don't strictly speaking "infect" us but live in our genetic material. Some endogenous viruses may have an even higher prevalence than HPV, yet they differ fundamentally from the free-living, or exogenous, viruses that represent our primary concern here. We will see these fascinating viruses again in chapter 7.

of women between the ages of fourteen and sixty in the United States, a whopping high rate for a virus. Rates are likely even higher in some parts of the world. Amazingly, the majority of sexually active humans on our planet, whether male or female, will get HPV at some point in their lives. The virus is made up of over two hundred different strains, all of which infect either skin or genital mucosa. Once the virus enters an individual, it generally stays active for years or even decades. Fortunately, the vast majority of HPV strains cause no problems for us. The few strains that do cause disease generally do so by causing cancer, the most important example being cervical cancer.* Luckily, most HPV simply spreads from individual to individual with little harm.

We're still aware of only a small percentage of all viruses that call humans their home. There may be viruses out there that infect even more people than HPV does; the work to identify all of the viruses that infect us has only just begun. Research over the last ten years has identified multiple previously unknown viruses circulating in humans that infect many individuals yet do not appear to cause any illness. The TT virus was named after the first individual to be infected, a Japanese man with the initials T.T. Very little research has been done yet on TTV, but it may be quite common in some locations. A report by one of my collaborators, Peter Simmonds, an excellent Scottish virologist, found prevalence rates ranging from 1.9 percent among Scottish blood donors to 83 percent among residents of The Gambia in Africa, a startlingly high range. Fortunately, TTV does not appear to be harmful.

GB virus is another recently identified and still largely unstudied virus present in many people. The virus got its name

---

* HPV and other viruses cause a great deal of global cancer burden and provide nontraditional approaches for preventing cancer that we'll revisit in detail in chapter 11.

from a surgeon, G. Barker; at the time, his hepatitis was mistakenly attributed to the virus.* I know from my own work how common both TTV and GBV are. Using very sensitive approaches to viral discovery, we frequently see these two—largely to our dismay, since they interfere with our ability to catch the dangerous culprits we're really looking for.

However common TTV and GBV are, they do not infect 100 percent of humans. So the literal Greek-derived definition of *pandemic* is probably an impossibility. The World Health Organization (WHO) has devised a six-stage classification of pandemics beginning with a class-one virus that infects just a few people and going on to a class-six pandemic, which occurs when infections have spread worldwide.

The WHO faced widespread criticism for labeling H1N1 a pandemic in 2009, but that's exactly what it was. H1N1 went from infecting only a few individuals in early 2009 to infecting people in every region of the world by the end of the same year. If that's not a pandemic, then I don't know what is. Whether or not we label a microbe that's spreading as a *pandemic* is unrelated to its deadliness. It's just a marker of its ability to spread. And as we discussed in chapter 1, the fact that H1N1 doesn't kill 50 percent of the people it infects (or even 1 percent for that matter) doesn't mean it won't kill millions of people or represent a massive threat.

In fact, from my perspective, it's possible that we could have a pandemic and not even notice it. If, for example, a symptomless virus like TTV or GBV were to enter into humans today and spread around the world, we probably wouldn't be able to tell. Most conventional systems to detect diseases only catch

* We'll also revisit GB virus in chapter 11. Some research suggests that not only is this virus harmless; under some circumstances, it might actually be good for you.

things that cause clear symptoms. A virus that didn't cause any immediate harm would likely be missed.

Of course, "immediate" isn't the same thing as "never." If a virus like HIV were to enter into humans today and spread globally, it wouldn't be detected for years, since major disease would occur sometime after initial infection. HIV causes only a relatively minor set of syndromes immediately, even though it starts to spread right away. AIDS, the major disease of HIV, doesn't emerge until years later. Since conventional methods for detecting new pandemics rely primarily on seeing symptoms, a virus that spreads silently would likely miss our radar, spreading to devastating levels before an alarm could be triggered.

Missing the next HIV would obviously be a catastrophic public health failure. Yet new viruses, even if likely to be completely harmless, like TTV and GBV, need to be monitored if they are moving quickly through the human population. As we saw in chapter 1, viruses can change. They can mutate. They can recombine with other viruses, mixing genetic material to create something new and deadly. If there's a new virus in humans and it's spreading globally, we need to know about it. The dividing line from spreading and benign to spreading and deadly is a potentially narrow one.

For our purposes, we'll define a pandemic as a new infectious agent that has spread to individuals on all continents (with the exception, of course, of Antarctica). One may counter that it would theoretically take only a dozen or so infected people to accomplish this—a few infected people per continent. That may be true, but it would be exceptionally rare for a microbe to spread so widely and infect so few individuals. And if it did manage to occur, even with twelve people, it would still represent a potent risk to all of us.

Defining precisely when a new spreading agent actually becomes a pandemic is less important for our objective here than understanding how pandemics are born. What I wanted to know when I began my research on pandemics was how something goes from being a strictly nonhuman infection to one spreading to humans on every continent.

In 2007 I worked with the aforementioned polymath biologist and geographer Jared Diamond and the tropical medicine expert Claire Panosian to develop a five-step classification system for understanding how an infectious agent living exclusively in animals can become an agent that spreads globally in humans. The system moves stepwise from agents that infect only animals (Category One) to agents that exclusively infect humans (Category Five).

Jared and I spent many afternoons pondering this process over extended writing sessions at his home in Los Angeles. During our lunch breaks, we'd stop writing to brainstorm, using thought experiments, how a virus might make this jump. We

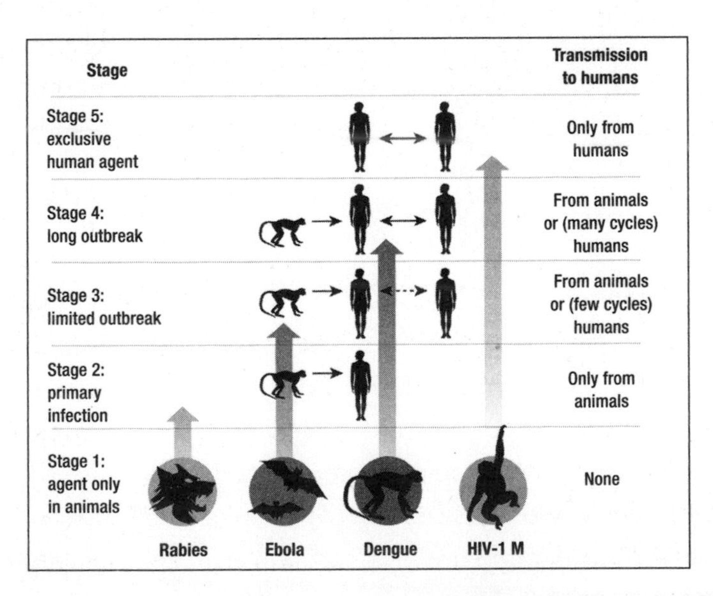

The five stages through which microbes of animals evolve to cause diseases confined to humans. (*Nature / Nathan Wolfe, Jared Diamond, Claire Panosian*)

came up with one fairly elaborate idea centered around the Diamonds' geriatric but much beloved pet rabbit, Baxter, and his invented disease—the dreaded Baxterpox. Even in our imaginary world, most human diseases have their start in animals.

Few of us now live on or near farms; fewer still live as hunter-gatherers surviving on wild plants and animals. We live in worlds filled with buildings and streets, where the dominant and notable forms of life are basically ourselves. Despite living on every continent with a population of seven billion individuals, we still represent a very restricted segment of the biological diversity on our planet.

As discussed in chapter 1, most of the diversity of life on our planet resides in the unseen world; in bacteria, archaea, and viruses. Despite our massive numbers and global reach, our human diversity pales in comparison. This is true even for our microbes. Most of the diversity in mammalian microbes resides in other animals, not humans. Some animals house greater microbial repertoires than others. For example, fruit bats are a notorious reservoir species. They often live in large colonies and are highly mobile "travelers" connecting multiple regions with high levels of biodiversity. On average a species of colonial fruit bat will have a greater diversity of microbes than, say, a two-toed sloth living a largely solitary life.

However you cut it, there are estimated to be over five thousand species of mammals on the planet and only one species of human. The diversity of microbes that *can* infect us from other mammals has and will always be substantially greater than the diversity of microbes that already *does* infect us. That's why we conceptualize the process as a pyramid, with the greatest diversity of microbes falling into Category One.

We've seen that most of the microbes with the potential to cause new human pandemics live in animals. Domestic animals certainly represent a threat, but as discussed earlier, most of what they had to contribute to the human microbial repertoire has already jumped over. Right now, the threat from domestic animals comes more from them acting as bridges to allow the movement of wild animal microbes into the human population. Moreover, while the actual number of living domestic animals is quite high, they represent a small percentage of the diversity of mammals since we've only domesticated a small percentage of all animals. Clearly, when it comes to new pandemics, wild animals are where the action is.

When I lived in Malaysia during the 1990s conducting my doctoral research, I spent time working with the accomplished parasitologists Janet Cox and Balbir Singh. Janet and Bal devised creative ways to detect malaria in blood that had been dried on small pieces of laboratory filter paper (it looks like plain but thick white paper). This technique made it easier to do field screening or specimen collection in remote locations. Since blood could be dried easily and stored at room temperature, this method did away with the logistics of having to keep a specimen cold in regions without electricity. Janet and Bal taught me how to use these lab techniques, and, with their sweet kids, Jas and Serena (now both college students!), introduced me to the amazing Malaysian state of Kelantan.

Kelantan is a small state on the border of Thailand that still adheres to Malaysian traditions that have disappeared in much of the now modern and economically booming country. Many of the people in Kelantan wear traditional Malaysian clothing, the official weekend is on Thursday and Friday, and there's not a drop of alcohol to drink (at least officially) in the majority of the state. The pace of life in Kelantan is more relaxed than almost anywhere I've visited in the now bustling, dynamic countries of Southeast Asia.

Among the fascinating sights to see in Kelantan was one

that held particular scientific interest to Bal, Jan, and me—coconut-picking macaques. In a unique practice, some coconut farmers in northern Malaysia and southern Thailand work with pig-tailed macaques, a species of Southeast Asian monkey, trained to climb palms and harvest coconuts. A well-trained animal can pick up to fifty coconuts an hour—quite an efficient farmhand.

One evening after dinner at their house, Bal told us of a report he'd heard of a man with a particularly devastating neurological disease, the symptoms of which suggested it was caused by a virus or other infectious agent. This man worked with the curious coconut-picking macaques.

The close and long-term relationship between macaques and their human handlers presented an ideal chance to study the first and second stages of the classification system I'd developed with Jared and Claire. We could study the microbes present in the animals and monitor them for any cross-species jumps into humans. Among the more interesting targets of our investigation would be the deadly herpes B virus.

Herpes B may not sound as if it should be particularly dreaded, but it's among the deadliest viruses a human can contract. Amazingly, the virus is almost completely benign among the macaques that sustain it. For the coconut-picking macaques, herpes B virus is just like herpes simplex is for a human, creating minor lesions that spread the virus through intimate contact like a bite or sex. Inconvenient for these monkeys, perhaps, but certainly not deadly. Yet when the virus crosses into humans, it causes severe neurological symptoms and invariably results in death. Transmission of the virus has been documented in a number of primate handlers in the West, including a sad case of a young woman working at the Yerkes Regional Primate Research Center in Atlanta who became infected after a captive macaque spit in her eye. At the time, no one had documented the infection occurring among the Kelantan coconut harvesters, despite the fact that they work with these animals daily and with far fewer protective measures in place.

For Bal, Janet, and me, studying the Kelantanese macaques and their handlers provided an interesting way to monitor the entry of these viruses into human populations and to witness the first step in the process of how potential pandemics are born. Yet, as with the case of rabies in young Jeremy Watkins, we did not expect to see these herpes B viruses go anywhere outside Category Two. They would remain as infections that had made the leap but didn't have the potential to spread in humans. While the victims might die from infection, the people infected by these monkeys would never go on to spread the virus to their families or others. Thus, no pandemic for herpes B. For that, we'd need a different kind of virus.

The interface we have with the animals in our world leads to a constant flow of microbes. Every day, millions of people are exposed to animal microbes. Some rare infections lead to death. Much more frequently, they are transient and benign infections, such as a bacteria from a pet dog or cat. The vast majority of these Category Two viral jumps represent dead ends from the perspective of a microbe: they infect a single individual, and that's that.

Sometimes, though, something unusual happens that is potentially pivotal for our species: a microbe that jumps over may have the capacity to move from one human to another human. If a microbe accomplishes this, it moves to a Category Three and toward becoming a pandemic.

In late August 2007 information began to trickle in to health authorities on an unidentified illness in a remote area of the Kasai-Occidental Providence in the Democratic Republic of Congo. The outbreak was centered around Luebo, a town of some historical importance as the last point that early twentieth-century steamers could navigate to on the Lua Lua River. The case reports listed a number of bad symptoms—fever, severe headache, vomiting, major abdominal pain, bloody diarrhea, and severe dehydration. The first recognized cases were on June 8, following the funerals of two village chiefs. Tellingly, the entire first group of individuals infected had assisted with the burials.

The symptoms and the connection to burials led the Congolese health authorities to consider the possibility of Ebola, a virus that spreads through direct contact with blood and body fluids, and they responded accordingly. The head of the Congolese team was Jean-Jacques Muyembe. Jean-Jacques is a professor and the director of the national biomedical research institute, the INRB. His bright laugh and mild-mannered demeanor belie the fact that he's had more experience dealing with viral hemorrhagic fevers* than perhaps any other single person in the world. I have fond memories of working with Jean-Jacques in a remote location in central DRC and watching him break into hysterics as he watched me devour a meal of pan-fried grub worms for dinner.

Jean-Jacques and his team called in long-standing collaborators, including Eric Leroy, a crack virologist who runs the only high containment bio-safety level four laboratory in central Africa that is capable of studying the world's deadliest viruses. Leroy, Muyembe, and colleagues at the CDC and from other groups like Médecines Sans Frontières (MSF) worked to contain the Luebo outbreak. They sequenced a small portion of the

* Viral hemorrhagic fevers, such as Lassa fever, Ebola, and others, all share severe symptoms, which include among other things a pronounced tendency toward swelling, broken capillaries, large-scale bleeding, low blood pressure, and shock.

virus's genetic information and discovered that it was, in fact, the Ebola virus.

Ebola hemorrhagic fever strikes fear in the hearts of people in the DRC and throughout the world. The Ebola virus kills quickly and dramatically. It also spreads. While the exact number of cases will never be known, the Luebo outbreak of 2007 probably infected around four hundred people. All of them were infected from a single virus that jumped from an animal into the first human victim and then subsequently spread. Around two-thirds of them died.

Part of the public fascination with Ebola relates to how little we know about something so deadly. The truth is that it largely remains a devastating and unsolved mystery.

What we do know about the Ebola virus is that it appears occasionally in humans. We know that it can enter into humans from multiple animal species. Leroy and his colleagues have identified the Ebola virus in a few species of bats, helping to pinpoint them as the likely reservoir. A range of studies also documents how Ebola affects gorillas, chimpanzees, and some species of forest antelope. We know that for now it's a Category Three microbe on the route to pandemics: it can infect and spread in humans, although not to the point of sustained transmission. Effectively, it's a virus with potential for localized outbreaks.

Together with Leroy and his colleagues, we looked in detail at the virus that caused the Luebo outbreak of 2007, as well as a smaller outbreak that occurred about a year later in December 2008 in exactly the same region of DRC. We found that the viruses that caused the two outbreaks were nearly identical and formed an entirely new type in the deadliest group of the Ebola viruses: the Zaire group.

That the Luebo outbreaks came from a new variant virus was significant. It meant that the depth of the genetic pool of viruses that could jump to us from animals was greater than we'd imagined. Now we understood that new versions of the Ebola virus had the potential to enter into humans, perhaps someone who hunted or butchered the meat of wild fruit bats. This meant that we probably haven't seen everything that the Ebola virus can serve up. For now, we classify Ebola as a Category Three agent, but our finding suggests that there are more undiscovered variants of Ebola out there that can cross into us. It's possible that a distinct and as yet unknown Ebola virus circulating in animals might have the potential to spread more broadly than any Ebola in the past.

Does the Ebola virus have the right stuff? Could it move higher in our pyramid classification system? From the perspective of a pandemic, all of the Ebola hemorrhagic fever outbreaks to date have been stillborn. They spread, but lucky for us, that spread remains limited.

Unlike the casual contact or airborne transmission of influenza, the majority of cases in the Ebola hemorrhagic fever outbreaks that have been studied resulted from intimate contact with the blood and body fluids of a very sick person. Generally, people become infected when preparing a previous victim for burial or when caring for the sick. Limited transmission makes broader, sustained spread less likely.

There are other disadvantages that the Ebola virus has in the microbial race to become a pandemic. The incredibly nasty symptoms of Ebola are both very specific and also coincide with its capacity to spread. Since few other viruses cause the dramatic symptoms of Ebola, it can be identified relatively quickly and the sick individuals can be isolated. Since it's the very sick people who spread the virus, isolation works to stop it. This is the approach that organizations like the CDC and MSF use to quell Ebola outbreaks: get in, isolate victims, stop contact with blood and body fluids. For the Ebola viruses that have emerged to date, it's a strategy that works. This kind of strategy often fails with more nimble viruses

• • •

In 1996 and 1997 quite another sort of outbreak occurred in the DRC. This outbreak lasted over a year, and while estimates vary, it likely hit over five hundred people. Like Ebola hemorrhagic fever, the cases began with fever, aches, and malaise. After a few days, rather than the bleeding characteristic of Ebola, patients developed a severe rash consisting of pustules all over the body, often first appearing on the face. The symptoms looked quite a bit like smallpox, perhaps the greatest scourge of human history. But that was impossible. Smallpox had been eradicated nearly twenty years earlier.

The cause of this outbreak was not smallpox, but it was a virus in the same group of viruses (the *Orthopoxvirus* genus) called monkeypox. Monkeypox has probably affected humans for ages, but it was only first recognized in 1970 during the smallpox eradication effort. Prior to that, any monkeypox cases were likely misdiagnosed as smallpox. While the ultimate ani-

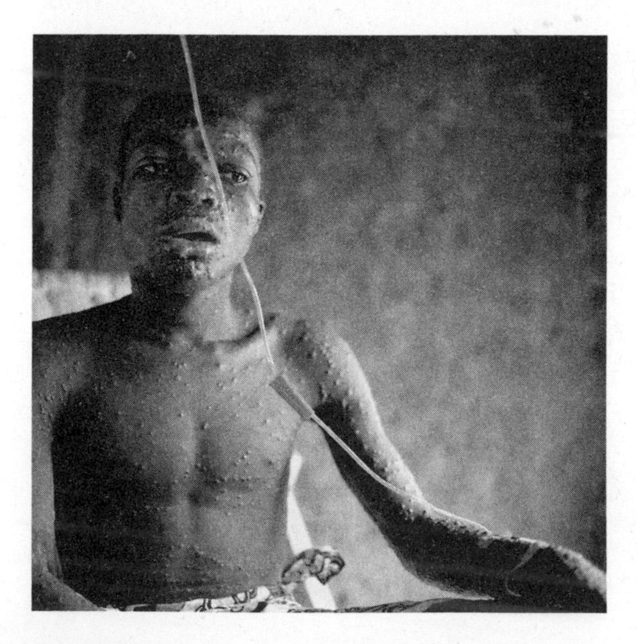

A young man with monkeypox. (*Lynn Johnson / National Geographic / Getty Images*)

mal reservoir for monkeypox remains unknown, it's almost certainly *not* a monkey, but rather a squirrel or other rodent. Because the virus can infect species of nonhuman primate, occasional human cases can result following contact with an infected monkey, hence the misnomer.

I've been working on monkeypox since 2005 with Anne Rimoin, an epidemiologist from UCLA, and her colleagues in the DRC, including Jean-Jacques Muyembe. Annie's spent much of the last ten years pushing deeper into the logistical nightmare of conducting high-quality surveillance for novel diseases like monkeypox in some of the most rural regions in the world. She manages to do it with flare. I've seen her touch up her eyeliner in the mirror of an off-road motorbike in a rural town in central DRC.

In 2007 we reported that monkeypox does not simply appear in outbreaks. The long-term work Annie and her colleagues did showed us that the virus should probably be considered endemic among humans—it is a permanent part of our world. Rather than follow the traditional method for investigating monkeypox

Dr Anne Rimoin in DRC. *(Prime Mulembakani)*

outbreaks, Annie and her team set up shop in regions that had known infections. Through constant monitoring, it became clear that there were monkeypox cases all year long. And the number of cases was growing.

In the final analysis it was just a matter of how hard you looked. During my visits to these sites, I've always seen cases of monkeypox. Some of these cases were the result of exposure to infected animals, but a number of them were the result of person-to-person transmission, the hallmark of a virus that's beginning to fully transition to a new host species.

You might wonder how such frightening cases of monkeypox could exist without the world being aware of them. The answer is that the region where we conducted this work is among the most remote in the world. Just to get to this area requires a chartered flight on a small plane or a three-week boat trip on tributaries of the Congo River that are only navigable during the rainy season. The setting is austere and beautiful, with very few roads. Most villages are linked together by simple footpaths. The research uses rugged off-road motorbikes traveling sometimes for as long as ten hours to get to the site of a case. Just dodging the chickens and pigs represents a major challenge.

Despite the incredible dedication and skill of our Congolese colleagues, the idea that the current meager resources devoted to health in the DRC could permit full coverage of a country four times the size of France is crazy. Yet this is one of the most important places in the world for the emergence of new viruses. Without a doubt, an interconnected world that doesn't invest in the infrastructure needed to monitor these viruses is doomed to fall victim to more epidemics.

Whether or not monkeypox has the potential to join the pantheon of our Category Four agents remains to be seen. Microbes that reach Category Four can live exclusively in humans while simultaneously continuing to live in animal reservoirs. Microbes in Category Four include dengue, discussed in chapter 4. Dengue maintains itself in human populations but also persists in a forest cycle spread by mosquitos among nonhuman primates.

Category Four agents represent the final step on the journey to become a human-specific microbe. They also present particular problems for public health. When scientists finally succeed at generating a vaccine for dengue, it will help countless people. But vaccination alone does not mean that we can eradicate dengue. Even if every single human were vaccinated, the fact that the virus can persist among monkeys in forests in Asia and Africa means that it will always have the potential to reenter human populations.

Monkeypox still ranks as a Category Three agent, but that could certainly change. Since our work in 2007, we've shown that the cases of monkeypox continue to grow in the DRC. Part of the explanation for this is that after smallpox was eradicated in 1979 the smallpox immunization program was stopped. As more and more nonimmunized, and therefore susceptible, children have been born into the population, the number of cases has steadily risen. And each additional case represents an opportunity for a unique monkeypox virus to jump or mutate. One of these may have the potential to spread and push monkeypox to the next level, which is why we keep tabs on this particular virus.

Only a handful of the microbes that have started on the path toward becoming exclusive human microbes have succeeded. The examples that have made it represent the mainstay of contemporary disease control. Viruses like HIV are generally considered to be present exclusively in humans, as are bacterial microbes like tuberculosis and parasites like malaria.* Yet it's often difficult to make the human-exclusivity call. Unless we have comprehensive data about the diseases of wildlife, it's hard to know if there may be a hidden reservoir of a supposedly exclusive human agent that could reenter human populations. And our understanding of the diversity of microbes in wild animals is still in its infancy. We know very little about what's out there.

Agents like human papilloma virus and herpes simplex virus almost certainly reside exclusively in humans, but they have likely been with us for millions of years. With an agent like HIV, we get into a gray area. Could the virus that seeded HIV a hundred or so years ago continue to live on in chimpanzees? Viruses very close to HIV have been found in chimpanzees, but we haven't sampled every chimpanzee in nature, so even closer relatives might still be out there. Similarly, given the diversity of malaria parasites we've seen in some of the African apes during recent studies, the possibility remains that some population of ape in some forest shares "human" malaria.

The question of reservoirs is an important one. We celebrated with great fanfare the eradication of smallpox in 1979. Eliminating that scourge from the human population was probably the greatest feat in public health history. Yet much remains unknown about how smallpox originated.

Smallpox appears to have first emerged during the domestication revolution. Evidence points to an origin in camels, which are infected with the closest known viral relative to smallpox,

---

* While the human malaria parasites are transmitted by mosquitoes from person to person, they still are considered "exclusively human agents." This is because they don't have another known animal reservoir and can't be sustained without *both* the mosquito and human part of their complex life cycles. If it's determined that another mammal can be infected with one of these parasites, then they'll be demoted to Category Four.

camelpox. Yet camels may very well have been a bridge host permitting the virus to jump from rodents, where most of the viruses like smallpox reside. If so, could there be a virus out there living in some North African, Middle Eastern, or central Asian rodent that's too close for comfort? A virus close enough to smallpox to reemerge and spread in humans? If so it might look a lot like monkeypox, and, like monkeypox, it might be largely missed.

For our purposes we should certainly consider smallpox to be one of our Category Five agents—a virus that made it to the point where it could live and survive exclusively in humans. And we should be proud of the herculean and successful effort to wipe it out.

Smallpox certainly had the right stuff. It probably killed more humans than any virus that has ever infected our species. Following the domestication revolution, the growing human populations and domestic animal populations (like camels) set the stage for the virus to gain a true foothold in our species.

We'll probably never know definitively what the first real pandemic was, but smallpox is a good candidate. It spread throughout the Old World after its likely camel origins but never made it to indigenous human populations in the New World on its own. When the Old World and New World collided at the onset of global travel some five hundred years ago, smallpox had the chance to make the jump, killing millions of the completely susceptible inhabitants of the Americas. That jump across continents positions it as the most likely candidate for the first real pandemic.

By the middle of the eighteenth century, smallpox had not only spread to every part of the world but had established itself just about everywhere, save for some island nations. And it killed. During the eighteenth century, it's estimated that smallpox killed around four hundred thousand people a year in Europe. The death rates elsewhere may have been even higher.

The human tendency to travel, to explore, and to conquer would accelerate dramatically over the five hundred years that would follow the discovery of the New World—and the coinciding smallpox pandemic. Global transportation networks would tie humans and animals together in a way that would accelerate the emergence of new viruses. These connections would result in a single, interconnected world—a world vulnerable to plague.

6

ONE WORLD

In 1998 scientists working independently in Australia and Central America announced that they were finding massive numbers of dead frogs in the forests where they worked. The large-scale die-off was especially dramatic. Global amphibian populations had been declining for some time, but these mounting frog deaths occurred in pristine habitats—places far less likely to have been exposed to toxic by-products of human cities or other man-made environmental threats. Field biologists and tourists alike witnessed the large numbers of dead frogs scattered about the forest floor. This was rare indeed since scavengers quickly eat dead animals. To see so many indicated that the predators already had their fill of free frogs and these were the leftovers. In fact, it was just the tip of the iceberg. A massive and unprecedented amphibian carnage was under way.

The expiring frogs all displayed similar and worrying symptoms. They became lethargic, their skin sloughed off, and they often lost their ability to right themselves if turned over. In the months that followed the first announcements, a number of possible explanations came forth—pollution, ultraviolet light, and disease among them. Yet the particular pattern of death

Frogs killed by the amphibian chytrid fungus. (*Joel Sartore / National Geographic / Getty Images*)

was most consistent with an infectious agent. Animal deaths spread in wavelike patterns from one location to the next suggesting the spread of a microbe, a contagion sweeping through the Central American and Australian frog world.

The solution to the mystery came in July 1998, when an international team of scientists reported the source of the frog disease. The team found evidence that a majority of the frog species succumbing to the die-offs were infected with a particular species of fungus. The fungus they identified was *Batrachochytrium dendrobatidis*, known more simply as the chytrid fungus (pronounced KIT-rid). They found evidence of chytrid, which had previously been seen exclusively in insects and on decaying vegetation, on a number of dead frogs. Tellingly, when they scraped the fungus from the dead and infected healthy laboratory tadpoles with it, they were able to re-create the fatal symptoms. The fungus was to blame.

Since the 1998 report, this fungus is now documented on all continents that have frog populations. It can survive at sea level but also wreaks havoc at altitudes up to twenty thousand feet. And it's a killer. In Latin America alone, chytrid fungus has been linked to extinction in 30 of the 113 species of the strikingly beautiful harlequin toads. Thirty species forever removed from the biological diversity of our planet.

While the spread and devastation of chytrid has now been well documented, much about it remains unknown. The large-scale declines in amphibian populations predated the emergence of the fungus, so it is not the only problem that is devastating global frog populations, but it's definitely among them. Another key factor is the steady decline in available frog habitat as the human footprint has increased over the last hundred years.

The questions of where the fungus originated and how it spreads are largely outstanding. Work done on archived specimens from South Africa shows that the fungus has infected African frogs since at least the 1930s, decades before it hit any other continent. This points to an African origin. Yet at some time, the fungus spread and did so quite effectively. How did it manage to get so cosmopolitan so quickly?

One possibility is the exportation of frogs. The researchers who discovered the early evidence of chytrid in South Africa also noted that some of the species of the frogs infected were commonly used in human pregnancy tests. When injected by lab technicians with urine from pregnant women, African clawed frogs (*Xenopus laevis*) ovulate—which made for an early, if significantly more cumbersome, version of the common pregnancy dipsticks used today! Following the discovery of this human pregnancy test in the early 1930s, thousands of these frogs were transported internationally for this purpose. Perhaps they took chytrid fungus with them.

But *Xenopus* was likely not alone in causing the global spread; since one stage of the fungus's life cycle actively spread in water, that was also a probable factor. Human movement almost certainly played a role as well. Our shoes and boots are at least partially to blame. This small fungus, wanted in the deaths of frogs worldwide, hijacked us.

The chytrid fungus has resulted in global frog deaths and in some cases extinction of entire frog species, a tragic loss for wildlife on our planet. In a 2007 paper, Lee Berger, one of the researchers who first identified the chytrid fungus, used language uncommon in conservative scientific journal articles when he wrote, "The impact of [chytrid fungus] on frogs is the most spectacular loss of vertebrate biodiversity due to disease in recorded history."

What happened with the chytrid fungus also gives us important clues to a larger phenomenon that affects much more than just amphibians. Over the past few hundred years, humans have constructed a radically interconnected world—a world in which frogs living in one place are shipped to locations where they've never previously existed, and one where humans can literally have their boots in the mud of Australia one day and in the rivers of the Amazon the next. This radically mobile world gives infectious agents like chytrid a truly global stage on which to act. We no longer live on a planet where pockets of life persist for centuries without contact with others. We now live on a microbially unified planet. For better or worse, it's *one world*.

How did we get to this point? For the vast majority of our history as living organisms on this planet, we had incredibly limited capacity to move. Many organisms can move themselves over short distances. Single-celled organisms like bacteria have small whiplike tails, or flagella, that allow them to move, but despite their molecular-scale efficiency, flagella will never push their owners far. Plants and fungi have the potential to move passively by creating seeds or spores blown by the wind. They also have adopted methods that co-opt animals to help them move, which explains the existence of fruit and the spores of fungi like chytrid. Nevertheless, precious few forms of terrestrial life regularly travel more than a few miles in the course of their lives.

Among the wonderful exceptions to the largely static life on Earth is the coconut palm. The seeds of the coconut palm (i.e., coconuts), like a number of other *drift seeds*, evolved buoyancy and water resistance, permitting them to travel vast distances through ocean currents. Among animals, some species of bats and birds are masters of space. The best example might be the Arctic tern, perhaps the most mobile species on Earth outside of our own. The tern flies from its breeding grounds in the Arctic to the Antarctic and then back again each and every year of its life. A famous tern chick was tagged on the Farne Islands in the UK near the time it was born in the summer of 1982. When it was found in Melbourne, Australia, in October of the same year, it had managed a twelve-thousand-mile journey in the first few months of life! It's been estimated that these amazing birds, which can live over twenty years, will travel about one and a half million miles in their lifetimes. It would take a full-time commercial jet pilot, flying at the maximum FAA permitted effort, nearly five years to cover the same distance.

Yet despite their wings, most bird and bat species actually live their lives quite close to where they're born. Only a few, like the Arctic tern, have evolved to regularly move great distances. Highly mobile species, whether bird, bat, or human, particularly the ones that live in large colonies, are of particular interest for the maintenance and spread of microbes. Among primates, only humans have the potential to move themselves great distances during a single lifetime, let alone in a few days. That's not to say that other primates simply stay put. Almost all species of primates move every day in their search for food, and young adults routinely move from one area to another before mating. Yet whether primate or bird, nothing on the planet—certainly nothing outside of the sea—matches humans in our capacity to move long distances quickly. The human potential to move, which now includes traveling to the moon, is unique and unprecedented in the history of life on our planet. But it comes with consequences.

Humans started globetrotting in earnest millions of years ago using our own two feet. Bipedalism gave us an advantage over our ape cousins in terms of our capacity to wander. And, as discussed in chapter 3, it had consequences for how we interact with the microbes in our environment. Yet our capacity to negotiate the globe in the amazing way we do now started with our use of boats.

The earliest clear archaeological evidence of boats dates to around ten thousand years ago. Found in the Netherlands and France, these boats (which might be better called rafts since they were made by binding logs together) were probably used primarily in fresh water. The first evidence of sea-going boats comes from a group of British and Kuwaiti archaeologists, who in 2002 reported finding a seven-thousand-year-old vessel that undoubtedly was used at sea. The archaeologists made their discovery at the Neolithic site of Subiya in Kuwait. Stored in the remnants of a stone building, the boat consisted of reeds and tar. Most strikingly, the bits of boat had barnacles attached to the tar, indicating that it was definitely used in the sea.

Employing genetics and geography, we can get a much earlier estimate for the first use of seafaring boats. The indigenous people of Australia and Papua New Guinea provide perhaps the best case for this. By comparing the genes of the Australasian people with other humans throughout the world, we can conclude that people reached Australia at least fifty thousand years ago.

During this time, our planet was a relatively cold place—it was the peak of an ice age. Since more of the Earth's water was locked up in ice, the sea level was lower, revealing pieces of land that connected what are currently islands. Many of the islands in the Indonesian archipelago were joined by these so-called land bridges.

Despite the land bridges that ice ages expose, we know that no one walked all the way to Australia. In particular, the deep-water channel between Bali and Lombok in present-day Indonesia, a channel around thirty-five kilometers long, would have required boats to navigate. So we can infer that these early populations also used at least some form of sea transport.

We know very little about these early Australian settlers, although we know that they traveled at a time before animal domestication so certainly didn't move with animals in tow. Nevertheless, their movements impacted how they related with microbes. When they first crossed from Bali to Lombok, they encountered a completely novel set of animals.

The channel between Bali and Lombok lies squarely on Wallace's Line, the famous geographic divide named after the nineteenth-century British biologist Alfred Russel Wallace who, along with Charles Darwin, codiscovered natural selection.* While the distance between Bali and Lombok was no greater than that between many of the waterways separating the hundreds of islands along the Indonesian archipelago, Wallace noted that animal populations on either side of the channel differed extensively. And while he didn't have the precise models for ice age water levels that we have today, he surmised that this biological divide existed because Bali and Lombok were never connected by a land bridge, something we now know to be true.

Like humans, other animals take advantage of land bridges, but unlike these earlier settlers who had boats, the animal

* Wallace led a fascinating life. Rather than working from a cushy boat like his contemporary and natural selection codiscoverer Darwin, he traveled on the cheap, selling specimens along the way to fund his expeditions. An excellent scientific biography of him as well as an accessible but detailed discussion of his findings in the Indonesian archipelago can be found in David Quammen's book, *The Song of the Dodo*.

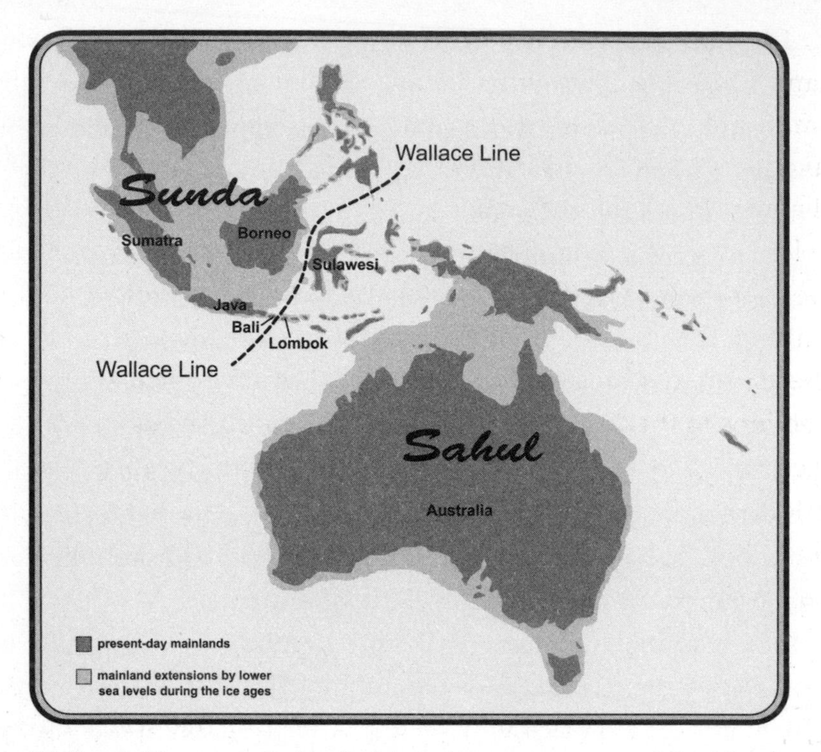

Wallace's Line, and the landbridges that once connected the islands on either side of it. (*Dusty Deyo*)

populations that couldn't fly long distances were largely stuck on one or another side of this deep-water barrier. When the first explorers left Asia for the Australasian continent, making the thirty-five-kilometer hop from Bali to Lombok, they took a fairly short trip by boat but a huge leap for primates. When they crossed this divide, these early explorers entered a world that had never seen monkeys or apes before. They also encountered completely new microbes.

These early settlers would have been hit with novel diseases from Australasian animals and their microbes, infectious agents that had never seen a primate before. Yet the impact of these agents for the human populations as a whole was likely limited, since the small population sizes of the settlers wouldn't have been able to sustain many kinds of agents.

It's hard to know exactly what the first trips across Wallace's Line were like. They may have been colonization events with small groups that were then completely cut off. Perhaps more likely they were short initial forays into new lands, followed by the establishment of temporary outposts, much as we consider colonizing the moon. The actual way in which the new lands were colonized would have played an important role in determining the flow of microbes in either direction. And while these first Australasian humans almost certainly had some connections to the "mainlanders" they left behind on Bali, that contact may have been very infrequent. Yet some new Australasian infections that had the potential for long-lasting human infection could very well have made their first forays into human populations on the Asian side of the divide.

The use of boats to visit new lands would continue with increasing frequency over the forty or so thousand years following this first colonization of Australasia. We have much better knowledge of what later trips were like and how they connected microbially distant lands. Perhaps the peak of boating-based colonization before modern times occurred among the Polynesian populations of the South Pacific.

Among these Polynesian journeys, probably the most incredible was the first discovery of Hawaii, over two thousand years ago.* For the first lucky settlers, finding this island would have been truly like finding a needle in a haystack. To give a sense

* The Polynesians had incredible navigation skills, and though their boats were simple, they were highly seaworthy. At a moment in history when boats in the West rarely went beyond the line of sight with land, Polynesians managed to negotiate huge swaths of the world's largest ocean. They fabricated their ships from two canoes, each dug out from tree trunks, which were lashed to each other with crossbeam planks to form a deck. They used coconut fibers and sap to seal seams.

of scale, the largest island of the Hawaiian archipelago, also named Hawaii, has a diameter of around a hundred miles. And the Southern Marquesas, whose inhabitants were the most likely first colonizers of Hawaii, are some five thousand miles away. To imagine what it would have been like to hit the mark, imagine we blindfolded an Olympic archer, then spun him around and asked him to hit his target—the ratios are about the same. One can only imagine how many boats (and their inhabitants) were lost before the fortunate finally made it.

On their long trips, the Polynesians probably lived largely on caught fish and rainwater. Yet they traveled with a veritable biological menagerie. They brought along sweet potatoes, breadfruit, bananas, sugarcane, and yams. They also traveled with pigs, dogs, chickens, and probably (unintentionally) rats. Having all of these domesticated species meant that the flotillas carried not only life support for the Polynesian explorers, but also minirepositories of microbes, which would spread and mix with local microbes in the places that they discovered.

The boat journeys of the Polynesians, as remarkable as they were for their time, pale in comparison to the global shipping that emerged in the fifteenth and sixteenth centuries. By the time Europeans reached the New World, in the late fifteenth century, thousands of massive sailing ships were plying the waters of the Atlantic and Indian Oceans and the Mediterranean Sea, moving people, animals, and goods back and forth between the countries of the Old World.

The impact of smallpox on New World populations is the most dramatic known example of the way that the connections formed by shipping can influence the spread of microbes. Some estimates suggest that as many as 90 percent of the people living in the Aztec, Maya, and Inca civilizations were killed by smallpox brought by boats during European colonization, a massive and devastating carnage. And smallpox was only one of many microbes that spread along the shipping routes of this time.

Each of the major transportation advances would alter connectivity between populations, and each would have their own impact on the spread of new microbes. The exclusivity of ships as a means for long-distance transport would not hold out forever. The use of roads, rail, and air provided new connections and routes for the movement of humans and animals as well as their microbes. For microbes, the transportation revolution was really a connectivity revolution. These technologies created links that forever changed the nature of human infectious diseases, including, critically, how efficiently they spread.

The use of roads of some sort or another is an ancient practice, far predating the use of water as a medium for transportation. Chimpanzees and bonobos both create and use forest trails to help them move through their territories. I learned this firsthand while studying wild chimpanzees in the Kibale Forest National Park in southwest Uganda. Richard Wrangham, the Harvard professor who introduced me to this work, used these trails to help observe chimpanzees.

Wrangham had done his doctoral work at the Gombe Stream site in Tanzania that Jane Goodall established. He'd critiqued some of the findings from Gombe because the chimpanzees there were habituated by provisioning—to get the wild chimpanzees comfortable with human researchers, the animals were fed large amounts of banana and sugarcane. Wrangham felt that provisioning changed some of the subtle chimpanzee behaviors, so when he started his own site in Kibale, he habituated the animals the hard way—by having his teams follow them until the apes effectively gave up and no longer ran away. He did this by essentially enhancing and extending the natural pathways that they moved along.*

The art of actual road building began in earnest around five to six thousand years ago when cultures throughout the Old World started using stone, logs, and later brick to enable the movement of people, animals, and cargo. The first modern roads followed in the late eighteenth and nineteenth centuries in France and the United Kingdom. These roads used multiple layers, drainage, and eventually cement to make permanent structures permitting regular movement throughout the year.

The rate at which modern roads have spread throughout the world has not been consistent, of course. Some regions in Europe and North America have roads reaching most human populations, while some regions where I work in central Africa have virtually no road access. Clearly, as roads enter into new regions, they bring both positive and negative effects. They are among the top priorities for many rural communities since they provide access to markets and health care, but from the perspective of global disease control, they are double-edged swords.

HIV is among the most notable example of the impact that road proliferation has had on the movement of microbes. In a series of fascinating studies, the HIV geneticist Francine McCutchan, whose lab I worked in at Walter Reed Army Institute of Research (WRAIR), and her colleagues at the Rakai and Mbeya sites in east Africa have examined the role that roads have played in the spread of HIV, demonstrating that proximity to roads increases a person's risk of acquiring HIV. As people have more access to roads, they have a higher chance of getting infected because roads spread people, and people spread HIV.

* Sue Savage-Rumbaugh, a primatologist from Georgia State University, has reported that bonobos go so far as to leave trail markers to help other group members find their way when conditions don't permit them to follow each other using footprints.

Other than sex workers, the highest occupational risk for acquiring HIV in sub-Saharan Africa is being a truck driver. McCutchan and her colleagues have shown that the genetic complexity of HIV is greater among individuals who have increased access to roads. Roads provide the mechanism for different types of HIV to encounter one another, in a single coinfected individual, and swap genetic information. But roads do more than just help established viruses spread. Roads and other forms of transport can also help to ignite pandemics.

One of the most stubbornly lingering public misconceptions is that we don't know how HIV originated. In fact, our understanding of the origins of HIV is more advanced than our understanding of the origins of probably any other major human virus. As we saw in chapter 2, the pandemic form of HIV is a chimpanzee virus that crossed into humans.* There is no debate within the scientific community on this point. The cumulative evidence with regard to how it originally entered into humans is also increasingly unequivocal. It was almost certainly through contact with chimpanzee blood during the hunting and butchering of chimpanzees. We'll delve further into this in chapter 9 when we discuss the work my colleagues and I have done with central African hunters.

Perhaps the only lingering debate about HIV origins is how it originally spread from the first infected hunter and why it took so long for the medical community to discover it. The earliest historical HIV samples date from 1959 and 1960, twenty years

* As discussed in chapter 2, HIV is a hybrid virus consisting of parts of two monkey viruses that chimpanzees acquired, almost certainly through the hunting of these monkeys. Note: There are multiple HIV viruses that have entered into humans (i.e., HIV-1M, HIV-1N, HIV-2, etc.). Here, when I refer to HIV, I mean exclusively HIV-1M, the dominant pandemic virus that is responsible for over 99 percent of human cases.

before AIDS was even recognized as a disease. In an amazing piece of viral detective work, evolutionary virologist Michael Worobey and his colleagues managed to analyze a virus from a specimen of lymph node from a woman in Leopoldville, Congo (now Kinshasa, DRC).

The lymph node had been embedded in wax for over forty-five years. By comparing the genetic sequence of the virus they found in the specimen with other strains from humans and chimpanzees, they were able to attach rough dates for the first ancestor of the human virus. While the genetic techniques they used cannot pinpoint dates closer than a few decades, they concluded that the virus split from the lineage sometime around 1900 and certainly before 1930. They also concluded that by the time that the woman in Leopoldville became infected with HIV in 1959 there was already a significant amount of genetic diversity of HIV in Kinshasa, suggesting that the epidemic had already established itself there.

The fact that HIV goes back to 1959, let alone 1900, provides some serious challenges to the medical community. One of the central questions is this: if it was in human populations in the early twentieth century and already constituted at least a localized epidemic in Kinshasa by 1959, why did it take us until 1980 to identify the epidemic? Another key question is what special conditions were present that permitted the virus to start taking off in the middle of the twentieth century?

A number of changes occurred in francophone central Africa, the region where HIV-1 originated, leading up to the period in the 1950s when those first precious samples were taken. The anthropologist Jim Moore and his colleagues at the University of California, San Diego, put together some of the key events in a 2000 paper, the majority of which focused on how easier means of travel influenced virus proliferation. In 1892 steamship service began from Kinshasa to Kisangani in the very heart of the central African forest. The steamship service connected populations that had been largely separated, creating the potential for viruses that previously might have gone extinct in local isolated populations to reach the growing urban centers. In addition, the French initiated the construction of railroads, which, like shipping and road lines, connect populations. This produced another mechanism for viruses to spread from remote regions to urban centers, effectively providing a larger population size of hosts for a spreading virus.

In addition to the connectivity provided by new steam, rail, and road lines, the construction of railroads and other large infrastructure projects led to cultural changes that also had an important impact. Large groups of men were conscripted, often forcefully, to build railroads. Moore and his colleagues note that the labor camps were populated mostly by men, a condition that dramatically favors transmission of sexually transmitted viruses like HIV. Together, the shipping and rail routes and the factors surrounding their construction must have played a role in the early transmission and spread of HIV.

As dramatic as the road, rail, and shipping revolutions were for the transmission of microbes, an entirely new form of transport would add another layer of speed. On December 17, 1903, in Kitty Hawk, North Carolina, a site chosen for its regular breeze and soft sandy landing areas, the Wright Brothers made the first sustained, controlled, and powered flight. Some fifty years later the first commercial jet flew between London and Johannesburg. By the 1960s, the age of jet travel was here to stay.

Airplanes link populations in an immediate way, which allows the transmission of microbes to occur even more quickly. Microbes differ from each other in terms of their *latent period,*

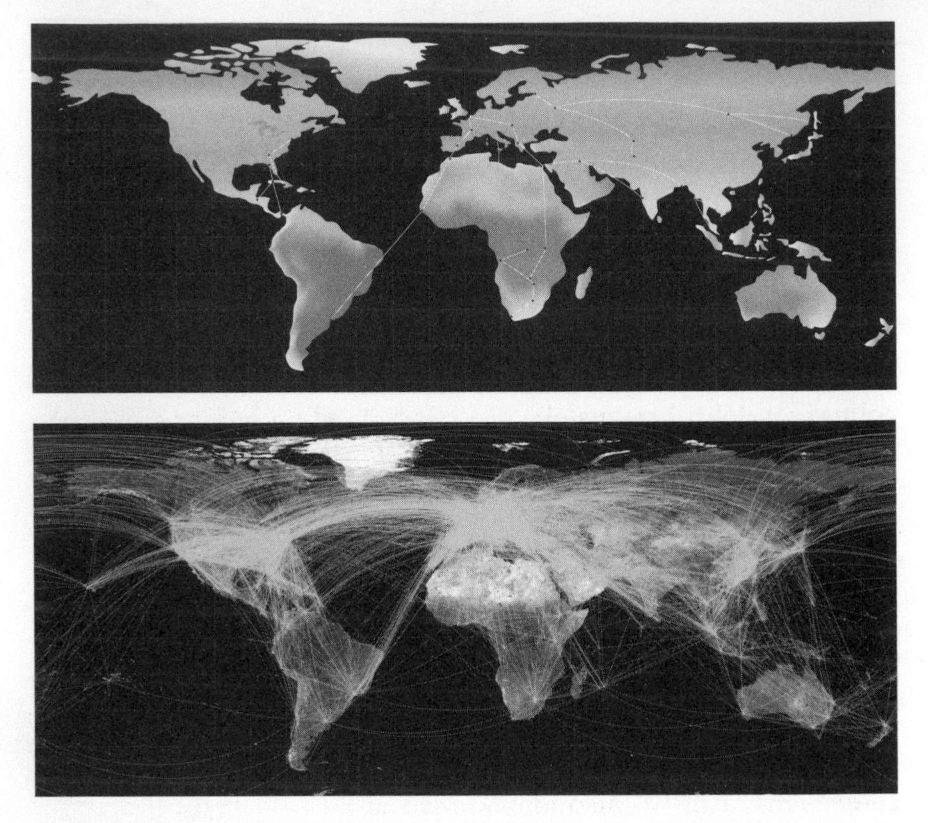

Top: World air traffic, 1933; Bottom: World air traffic, 2010. (*T: Dusty Deyo; B: OpenFlights.org*)

the period of time between when an individual is exposed and when they become infectious or capable of transmitting the agent to others.* Almost no microbes that we know of have

* The *latent period* differs subtly but in an important way from the *incubation period* for some microbes. Where the latent period refers to the time between exposure and infectiousness, the incubation period refers to the time between exposure and the first signs of disease. In the case of HIV, for example, infected individuals become contagious within the first few weeks after exposure, yet at this point they experience only generic symptoms like fever

latent periods of less than a day or so, but many have latent periods of a week or more. The immediacy of air travel means that even microbes with very short latent periods can spread effectively. In contrast, if a person infected with an agent that had a very short latent period were to board a ship, unless the ship had hundreds of individuals the virus could infect, it would go extinct before the ship made land.

Commercial air flights alter in fundamental ways how epidemic disease spreads. In a fascinating paper from 2006, my colleagues John Brownstein and Clark Freifeld of Harvard, one of the new academic breed of *digital epidemiologists*, found creative ways to use existing data to show just how much impact air travel has on the spread of influenza. John and his colleagues analyzed seasonal influenza data from 1996 to 2005 and compared it with patterns of air travel. They found that the volume of domestic air travel predicts the rate of spread of influenza in the United States. Interestingly, the November travel peak around Thanksgiving appears to be of particular importance. International travel also plays a vital role. When the number of international travelers is lower, the peak of the influenza season comes later—because when there are fewer travelers, it takes longer for the virus to spread. Perhaps most strikingly the researchers were able to see the impact of the terrorist attacks of September 2001 on influenza. The travel ban led to a delayed influenza season. The striking effect was not seen in France, which did not enact the ban, providing an excellent control.

During the past few centuries the ease of movement has increased dramatically throughout the world. The rail, road,

---

and rash. Most cases of HIV transmission actually occur during this acute infection period rather than after the incubation period for AIDS itself, which is generally some years later.

sea, and air revolutions have all permitted humans and animals to move more quickly and efficiently both within continents as well as between them. The transportation revolution has created interconnectivity unprecedented in the history of life on our planet. It is estimated that we now have over fifty thousand airports, twenty million miles of roads, seven hundred thousand miles of train tracks, and hundreds of thousands of ships and boats in the oceans at all times.

The connectivity revolution we've experienced has fundamentally changed the ways that animal and human microbes move around our planet. It has radically increased the speed at which microbes can travel. It has brought populations together, allowing agents that couldn't previously sustain themselves with low population numbers to flourish.

As we'll see in chapter 8, it has also permitted completely novel diseases to emerge and frightening animal viruses to extend their ranges. These technologies have created a single interconnected world—a giant microbial mixing vessel for infectious agents that previously stayed separate and stayed put. The new microbial mixing vessel that our planet has become has forever altered the way in which we'll experience epidemics. It has truly helped usher us into the pandemic age.

► 7 ◄

THE INTIMATE SPECIES

On February 2, 1921, Englishman Arthur Evelyn Liardet went into surgery. Liardet's symptoms were typical, but the surgery was not. At the time of the operation, Liardet was seventy-five years old and complained of decreases in his physical and mental energy. He had lost most of his hair and had developed wrinkle lines on his face. In short, he was growing old.

Some years before that chilly day in 1921, Liardet had written to an up-and-coming Russian surgeon practicing in Paris and offered his body for a unique procedure. The surgeon, Serge Voronoff, claimed to offer nothing other than total bodily rejuvenation—the elixir of life.

Serge Abrahamovitch Voronoff was born in Russia in 1866. At the age of eighteen, he immigrated to France, where he studied medicine under the Nobel laureate Alexis Carrel. Carrel had won his Nobel Prize in 1912 for surgical work on blood vessels as well as the new methods of transplantation of both blood vessels and whole organs. Carrel taught Voronoff surgery, undoubtedly impressing upon him the excitement of science and the potential for discovery, particularly surrounding the revolutionary new techniques of organ transplantation. In doing so, he launched the career of the young, ambitious, six-foot-four Voronoff, described in reports as magnetic and imaginative.

Following his studies with Carrel, Voronoff worked in Egypt for the Egyptian king. Voronoff soon became fascinated with the eunuchs that were part of the king's harem. In particular, he noted that the castration they received seemed to increase the speed at which the eunuchs aged. This observation was the beginning of Voronoff's obsession with a surgical answer to aging. Likely inspired by the pioneering work of his mentor and the excitement of the new surgical techniques, Voronoff began to dabble in experimental transplantation. But he went beyond the techniques that his mentor had perfected. In early experiments Voronoff transplanted the testicles of a lamb into an old ram, claiming that the transplant served to thicken the ram's wool and increase its sex drive. These early studies foreshadowed the work that would follow.

Some years later, on that cold February day in Paris, Liardet became one of Voronoff's early human experiments. Before Liardet was wheeled into the operating room, a chimpanzee had been anesthetized using a specialized "anesthetizing box" developed by Voronoff. The box served to protect the technicians from the massive and potentially violent male chimpanzees, who would have certainly reacted strongly to what was coming. Then Liardet was wheeled in on a gurney and placed alongside the chimpanzee. The surgeons carefully removed a testicle from the chimpanzee, cut it into thin slices, and grafted pieces onto the testes of Liardet.

The procedure, known in its day as the monkey gland operation, would go on to become remarkably popular. By 1923 forty-three men had received testicles from nonhuman primates, and by the end of Voronoff's career, that number reached the thousands. Although Voronoff had inherited a fortune as an heir

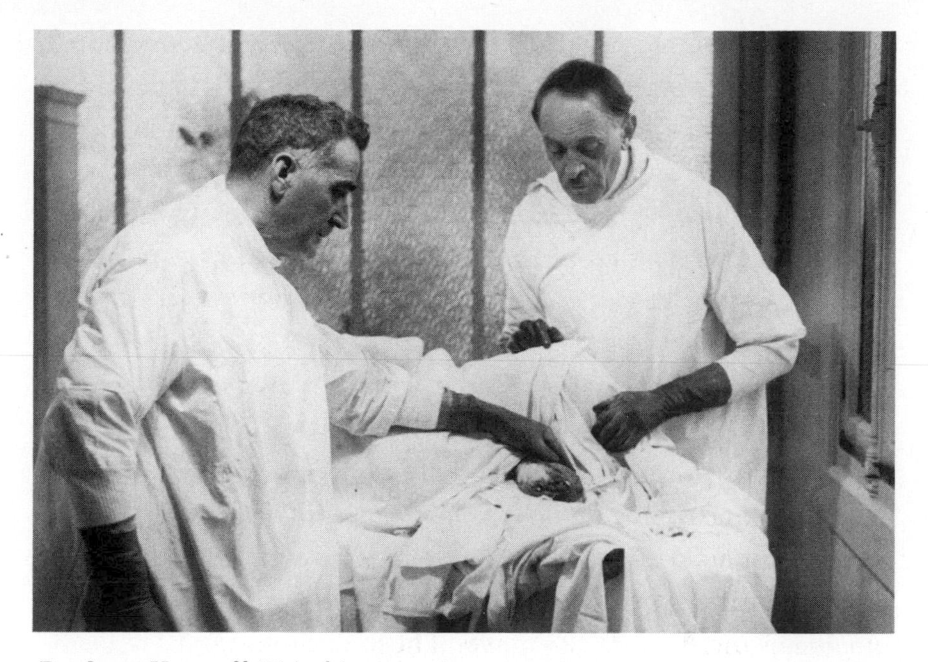

Dr. Serge Voronoff (R) in his operating room. (© *Bettmann / Corbis*)

to a vodka manufacturer, he made more money operating on many of the most important men of his day. Unverified but highly specific accounts point to Anatole France, a Nobel Prize–winning French poet as one of his patients. Less reliable rumors suggest Pablo Picasso might also have gone under Voronoff's knife.

But to what end? Many of Voronoff's patients swore by the procedure. Liardet himself claimed to a *New York Times* reporter in 1922 that the procedure had been a huge success. He showed the reporter his strong biceps while his wife nodded knowingly by his side. Though the successes claimed by Voronoff and his patients were certainly exaggerated, the underlying logic of the procedures remains open to at least some scientific debate. Voronoff himself, as well as his procedure, eventually fell out of favor with the scientific community. By the time of his death in the early 1950s, most considered him a quack, perhaps in part because of thc cxtremes to which he went. In the most dramatic of his experiments, he transplanted ovaries from a human woman into a female chimpanzee named Nora. He then attempted to inseminate her using human sperm!* Yet a 1991 editorial on Voronoff in the top British medical journal *The Lancet* concludes with the following words: "Maybe medical research councils should fund further research on monkey glands."

For our purposes, the utility of the chimpanzee testicle transplants as the Viagra of the roaring twenties, while interesting, is largely unimportant. What's important about the infamous monkey gland operation is that it provides one of the most striking examples of how medical technologies have incidentally served to link humans (and sometimes animals) in ways that create new bridges for the movement of microbes.

Given what we now know, the idea that we would consciously connect the microbial worlds of humans and chimpanzees as Voronoff did would be unfathomable and unforgivable. Although there's no straightforward way to confirm it in the absence of specimens, Voronoff's transplants almost certainly led to the transmission of potentially dangerous viruses into humans who received these tissues. Transplanting living tissue between very closely related animals eliminates all of the natural "barriers to entry" that microbes face, and remains one of the riskiest imaginable ways for a microbe to jump from one species to the next.

Yet Voronoff's work, while certainly extreme, did not exist in a vacuum. The explosion of medical technologies that has

* As noted in the book *Charlatan* by Pope Brock, which provides excellent background on Voronoff, "The only thing that [the insemination] produced was a novel, Félicien Champsaur's *Nora, la guenon devenue femme—Nora, the Monkey Turned Woman.*"

occurred over the past four hundred years has provided new kinds of microbial connections between individuals. Transfusion, transplantation, and injection, while some of the most critical tools for maintaining human health, have also contributed fundamentally to the transmission and emergence of pandemics. These technologies have connected us with one another's blood, organs, and other tissues in ways unprecedented in the history of life on our planet. They have served to make us, among other things, the *intimate species*.

Before we dissect the role of medical technology in connecting our bodies and facilitating microbial transmission, it's worth spending a moment to discuss this in the context of the benefits of these technologies. Injections and immunizations as well as transfusions and transplants are all technologies that have catapulted medicine into the modern age.

Without blood transfusions, huge numbers of hemophiliacs, trauma victims, and wounded soldiers would die. Transplants permit victims of leukemia, hepatitis, and severe burns to live a normal life. And it is nearly impossible to imagine a world without injections. The use of intravenous rehydrating fluids alone saves the lives of millions of malnourished children and victims of diarrheal disease each and every year. Injections also permit immunization, and to live in a world without immunizations is to live in a world where smallpox threatens our day-to-day lives. If smallpox had not been eliminated by immunization in the 1960s, it would probably be a worse plague today than it was then due to the same sort of hyperconnectivity discussed in chapter 6.

Examining the role that these medical technologies have played in the history of epidemics does not argue against their utility in maintaining the health of our species. Likewise, these arguments should not be interpreted as supporting the hypochondriacal fears of the anti-immunizationists, whose rhetoric has been soundly skewered in Michael Specter's important book, *Denialism*, which brings to a general audience years of research on the subject. Nevertheless, understanding the ways in which the historical use of these technologies has connected human populations is important in understanding why we are plagued with pandemics. Rather than discourage us from using live-saving technologies, they serve to highlight the need for maintaining vigilance in the way that we deploy them.

One of the clearest examples of how medical technology increases human microbial interconnectivity is our use of blood. Historically, humans (and other animals) rarely come into contact with one another's blood. For the vast majority of our history since the advent of hunting, we've had more contact with the blood and body fluids of other animals, through hunting and butchering. In the fifteenth century, that would all change.

The first generally accepted attempt at a "blood transfusion" was given to Pope Innocent VIII in 1492 and was described by the historian Stefano Infessura.* When the pope went into a coma, the pontiff's medical advisers sought to cure him by feeding him the blood of three ten-year-old boys. At the time, there were no intravenous techniques, and both the pope along with the three donors, who had been promised a ducat each, all died.†

Blood transfusions have advanced considerably since then.

* The account is described in a book on the papacy by the contemporary historian Peter De Rosa.

† Interestingly, the first documented intravenous blood transfusion was from an animal to human, rather than from one human to another. In June 1667, Dr. Jean-Baptiste Denys, the physician to none other than King Louis XIV of France, administered a transfusion of sheep blood into a fifteen-year-old boy. Nothing is known about the sheep, but we know the boy survived.

Today around eighty million units of blood are collected every year worldwide. Blood transfusions save countless lives. They also provide an entirely new form of connection between human populations. When a unit of blood is transfused from individual to individual, so too are the various viruses and other microbes present in that unit. The proliferation of blood transfusions creates a novel route for the movement of microbes, sometimes providing a new route for microbes that already move in other ways, as in the case of malaria. New connections made through medical technology can also provide transmission routes for agents that would otherwise never have the potential to spread among humans and can be another way in which an agent that we acquire from an animal can spread; otherwise, it might simply go extinct.

Blood transfusions are known to spread HIV and other retroviruses, as well as hepatitis B and C, parasites such as malaria, and Chagas disease. Even prions like variant Creutzfeld-Jakob (also known as mad cow disease), an infectious agent to which we'll return in the next chapter, can survive in blood bags, the plastic containers used to hold blood prior to transfusion.

The field of blood products has also advanced beyond single blood transfusions that go from one person to the next. In the case of hemophiliacs, individuals lack particular blood factors that would normally allow their blood to clot, a potentially life-threatening condition. In order to accumulate the missing blood factors in sufficient quantities to solve the problem, the appropriate components of blood are often pooled from tens of thousands of donors. The consequences for connectivity are substantial. A back-of-the-envelope calculation suggests that a person living with hemophilia A in a city like San Francisco will inject themselves with up to 7,500 doses of clotting factor VIII by the time that they're sixty. That means that this person will potentially have had contact with the blood-borne microbes of 2.5 million people during the course of their lifetime.

The good news is that many blood banks now screen for the usual suspects. Blood infected with HIV, for example, will not get through the donation process.* But this was not always the case. Perhaps not surprisingly, hemophiliacs who receive pooled blood products were among the first found to be infected with HIV in the early 1980s. In the United States alone, thousands of hemophiliacs were infected, and many of them died. Even now we can only screen for the microbes we know. Undiscovered viruses, of which there are certainly many, move around daily through blood products. And a new virus that entered into humans could easily spread within the blood supply before we'd have a chance to stop it.

In terms of sheer number of procedures, blood transfusions trump organ transplants by far. But while the number of blood transfusions will always be greater than the number of organ transplants, the movement of an organ is a substantially more dramatic biological event. Organ transplantation involves the movement of blood as well as a large amount of tissue, so any microbes present in the blood or the tissue being transplanted will move along with the organ into the recipient.

Organ donation risks the movement of all of the same usual suspects that will transmit during a blood transfusion and then some. For example, in the case of rabies, discussed in detail in chapter 5, I told you that rabies didn't transmit from human to human. To be slightly more exact, there have been no documented cases of natural transmission of rabies from human to human. There have, however, been a dozen or so well-documented cases of rabies moving from one person to

* Sadly, there is huge variation in the extent to which blood banks screen. Screening in developed countries is generally excellent, while in some parts of the world, it remains virtually nonexistent.

another. And every one of them has been due to a transplant with an infected organ.

The majority of these cases of rabies transmission have been due to cornea transplants, perhaps because the cornea is one of the only nervous system-related tissues currently transplanted; and rabies is primarily a virus of the central nervous system. In two fascinating cases, separate recipients from Texas and Germany received organs infected with rabies. In both cases the deaths of the donors were falsely attributed to drug overdoses, whose symptoms can sometimes mimic those of rabies. Amazingly, both donors appeared to have died of fulminant rabies without seeking medical care. It's frightening to imagine that they reached that point in the illness while moving about, conducting their normal day-to-day affairs.

Transplants can also transmit dormant stages of infectious diseases that can flare up later. One particular species of malaria, *Plasmodium vivax*, whose Siberian variant we discussed in chapter 1, has the capacity to stay latent, or dormant, in the liver. During latency, there are no symptoms of the disease and, similarly, no malaria parasites in the blood. But a liver transplant could potentially do the trick.

In one case in Germany, a twenty-year-old male who had emigrated from Cameroon died of a brain hemorrhage and donated his organs. Among the recipients was a sixty-two-year-old woman who needed the liver because of late-stage cirrhosis. A full month after her organ transplant she developed a high-grade fever, which was later diagnosed as *vivax* malaria and successfully treated. She had never visited a tropical or subtropical area in her lifetime. But her liver had.

There's another important rub to the problem of organ transplantation. While obtaining blood from donors is pretty simple, obtaining an organ is not. Rarely will people living in the developed world who need a blood transfusion *not* get one, but that is not the case with organ transplants. Currently in the United States there are approximately 110,000 people on the waiting list to receive organ transplants, and one of these people dies every ninety minutes.

The lack of organs available for transplant has caused surgeons to look for alternatives to human organs. Animals are an obvious choice. Clearly, the elective transplantation of thousands of chimpanzee testicles into men for the purposes of "rejuvenation" presents unacceptable risks. The real chance that a known or unknown virus could enter into humans from such a closely related animal as a chimpanzee would make the choice wrong even if the illness were life-threatening. But there are other animals out there. For example, the organs of an adult pig are approximately the size and weight of an adult human's, and while not without risk for cross-species jumps, the risks are smaller than those for a chimpanzee.

In July 2007, Joe Tector, a pioneering transplant surgeon at Indiana University School of Medicine, gave me a call. Tector had put together a team whose goal was to genetically engineer pigs with organs that would be less likely to be rejected by humans, a problem for surgeons to this day. Within five years he wanted to begin transplanting pig livers into humans, and he wanted to understand the risks.

He explained that he didn't want a rubber stamp scientist to simply tell him that what he was doing was safe. He was looking for someone who loved finding new viruses and was intent on discovering any possible risk associated with the work he was doing. He spoke eloquently about his patients and how many of the people on those waiting lists never received organs in time. He also spoke about the need to determine that what he planned to do wouldn't backfire, igniting a new pandemic in the human species. When Tector contacted me, I had already been interested for some years in xenotransplantation, the surgical term for the transplantation of organs from one species to another. By the time I got off the phone, I was hooked.

In the 1920s, following the time that Voronoff conducted his work in Paris, the field of xenotransplantation went into a forty-year lull during which there were no documented attempts. But in the 1960s work on xenotransplantation had a rebirth. New antibiotics and immunosuppressive drugs provided hope for the success of transplanting major animal organs into people who needed them. By dampening the immune system, the immunosuppressive drugs could address the frustration of organ rejection.

A series of high-profile operations brought major attention to the field through the 1980s. One involved the famous Baby Fae, a twelve-day-old baby girl born prematurely with a major heart disorder. She survived for eleven days on a baboon heart. Another operation created a news flurry around Jeff Getty, a thirty-eight-year-old man with AIDS. Getty received his diagnosis at a time when AIDS was still called "gay cancer." He went on to become a prominent activist who pushed for access to care for AIDS patients and participated in numerous experimental trials, including the one that led to his national notoriety. In the trial, he received a bone marrow transfusion from a baboon with the hope that the natural resistance that baboons have to AIDS would take hold in his body.

Getty's experimental therapy ultimately failed, but it brought with it a national debate about the potential that such transplants could transmit new and perhaps unknown viruses to humans. Certainly, a transplant from a closely related species like a baboon to a person with an already compromised immune system could be a recipe for disaster. A person with a highly weakened immune system, as occurs in late stages of AIDS, would provide an environment in which new viruses could better grow and adapt.* In the extreme, it could be a Petri dish for viral exploration into a new and foreign land.

Fortunately for the success of Tector's work, pigs are not as closely related to humans as baboons are. Yet they are still mammals. As with other mammals (including ourselves), they have many microbes that are still unknown. And some of them undoubtedly have the potential to jump species. The real question then becomes what are the viruses that *can* jump and can they then spread from person to person. The fact that one person might get a deadly virus is not the end of the world, particularly if the person was about to die of liver failure. The real risk is if that virus could spread.

In the small but active group that concern themselves with pig viruses, the agent that has provoked the most worry is PERV, the porcine endogenous retrovirus. Endogenous viruses like PERV are permanently integrated into the genetic material of their hosts. Yet from time to time they emerge from the genes and go on to infect cells and spread within the host's body. As part of the actual genomes of their hosts, endogenous viruses cannot currently be eliminated—hence the concern that they could reemerge in humans following a pig transplant.

The eminent CDC virologist Bill Switzer, whom I've worked closely with for the last ten years studying retroviruses, was one of the scientists to conduct the most comprehensive study on PERV in xenotransplant recipients. Bill and his colleagues studied specimens from 160 patients who had received pig tissues. Amazingly, they found evidence of pig cells continuing to live in about 15 percent of the recipients, even up to eight years after the transplant. Fortunately, they found no evidence of PERV.

---

* Patients with AIDS are not the only ones who are immunosuppressed. Transplant recipients commonly receive drugs that suppress the immune system in order to prevent organ rejection, which likely means that everyone receiving organs today, whether afflicted with AIDS or not, are at an increased risk of infection.

Whether PERV is the most important risk or not remains unknown. If it is, we may not have much to worry about. Through our studies with Tector and other colleagues, we hope to determine what else may be in those pig tissues and what risks those agents would pose. The decisions based on our research will not be easy ones. As we'll discuss further in chapter 9, even state of the art viral discovery right now does not permit us to definitively determine all of the microbes in any sample. Yet the costs of indecision are substantial. On one side are the transplant recipients who die each day waiting for an organ. On the other is a small but important risk of an epidemic in a much larger group. Is one life saved worth a species potentially plagued?

We've been sticking ourselves with needles for a long time. The first evidence of it comes from an unusual source—an iceman. On a sunny day in September 1991, two German tourists hiking in the Italian Alps came across a corpse. The corpse became known as Ötzi, after the valley in which he was discovered. Though initially thought to have died recently, we now know that Ötzi lived 5,300 years ago.

Among the amazing elements of this discovery is the fact that Ötzi had tattoos. In fact, this is the first evidence of tattoos in the world. Ötzi's tattoos were located on his lower back, ankles, and knee. X-rays of the mummy showed evidence that the tattoos were positioned over spots where Ötzi had likely experienced pain due to orthopedic maladies, leading some to speculate that the tattoos may have served as a kind of therapy.

Whatever his reasons for having them, Ötzi's tattoos, like any tattoo since, represent risks. Tattooing, like a needle stick or an injection, involves blood contact. And if the same implement is used multiple times on different individuals, it can provide a bridge on which microbes can hop hosts.

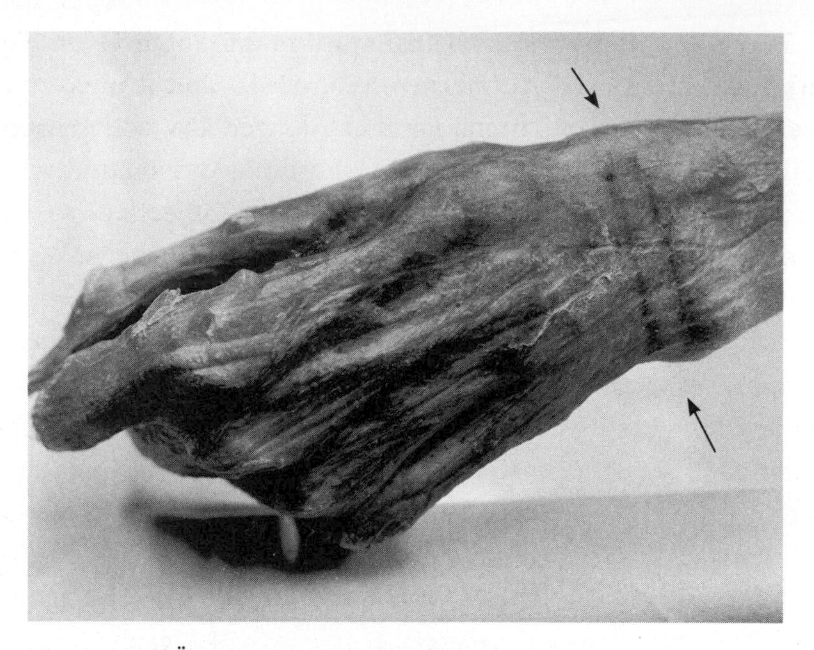

The wrist of Ötzi the Iceman, showing two of his numerous tattoos. (© *South Tyrol Museum of Archaeology,* www.iceman.it)

Whether for tattoos, medicines, or vaccines, improperly sterilized needles can play an important role in transmitting microbes. Widespread use of needles, as with blood transfusions, provides an entirely novel route for microbes to move around, allowing them to maintain themselves or spread effectively in humans in order to survive and thrive.

Perhaps the most remarkable microbe we know of in the postinjection age is hepatitis C virus. HCV is a critically important virus that infects over one hundred million people globally and more than three million new individuals each year. It also kills through liver cancer and cirrhosis, causing over eight thousand deaths per year in the United States alone. But it would likely kill precious few of those individuals if it weren't for needles.

There is still a great deal that's unknown about HCV. The virus itself was officially discovered in 1989, but it must have been in human populations for much longer. My collaborator, the prolific Oxford virologist Oliver Pybus, has made understanding this virus one of his many scientific objectives. Pybus utilizes the tools of evolutionary biology and learns more each year about viruses through computers than many others will in a lifetime of lab- or fieldwork. By using computer algorithms to compare genetic information from distinct viruses as well as mathematical modeling, Pybus has made some fascinating discoveries about HCV.

What we do know about HCV is that it's on the move. During the past hundred years, the virus has spread rapidly through blood transfusions, the use of unsterilized needles to deliver medicines, and through injection drug use. But genetic analyses by Pybus and others have shown that the virus is somewhere between five hundred and two thousand years old, so these contemporary technologies likely do not tell the whole story. Essentially, it seems there were places, most likely in Africa and Asia, where HCV existed on a much smaller scale prior to the massive expansion that needles and injections permitted.

Since HCV is not effectively transmitted sexually or by normal contact between people, other routes of transmission must somehow explain how it persisted many centuries ago. The virus can be transmitted from mother to offspring, but that too is an unlikely explanation since so-called vertical transmission is not particularly efficient. Certainly, cultural practices like circumcision, tattooing, ritual scarification, and acupuncture probably played a role. In an interesting twist, Pybus and his colleagues have used a combination of geographic information systems (to which we'll return in chapter 10) and mathematical models of disease spread to show that another possibility would be some kinds of blood-feeding insects. These insects could have contributed to historical transmission by acting as natural contaminated needle sticks and carrying virus-infected blood from one host to the next on their mouthparts.

Unsafe injection practices contributed to the spread of much more than HCV in the twentieth century. In a series of thoughtful articles, the Tulane virologist Preston Marx and his colleagues have argued that injections helped launch the HIV pandemic. Some mysteries still persist on the early spread of HIV. While the genetic data points to an early twentieth-century jump of the chimpanzee virus that would become HIV, understanding what sparked its true global spread in the 1960s and 1970s remains up for debate. For many scientists, the expanding air routes discussed in chapter 6 are sufficient to explain this phenomenon. But Marx and his colleagues add another potential cause.

The period that coincides with the global spread of HIV also coincided with a dramatic expansion in the availability of cheap injection systems. Prior to the 1950s, syringes were handmade and relatively expensive. But in 1950 machines began churning out glass and metal syringes, and in the 1960s disposable plastic syringes became available. Effective ways to inject drugs and vaccines contributed to the increased use of injections for medications and vaccines in the late nineteenth century. Often medical campaigns used the same unsterilized needle to vaccinate hundreds or more individuals at a time, setting up unique conditions that could potentially launch epidemics.* An individual hunter who had been infected with a virus from a hunted chimpanzee

* Marx and his colleagues argue that such multiple injections would naturally simulate the "serial passage" experiments done in laboratories with viruses. In these experiments, viruses are moved from animal to animal in a way that produces extensive opportunity for the accumulation of mutations that permit the virus to survive in a novel host.

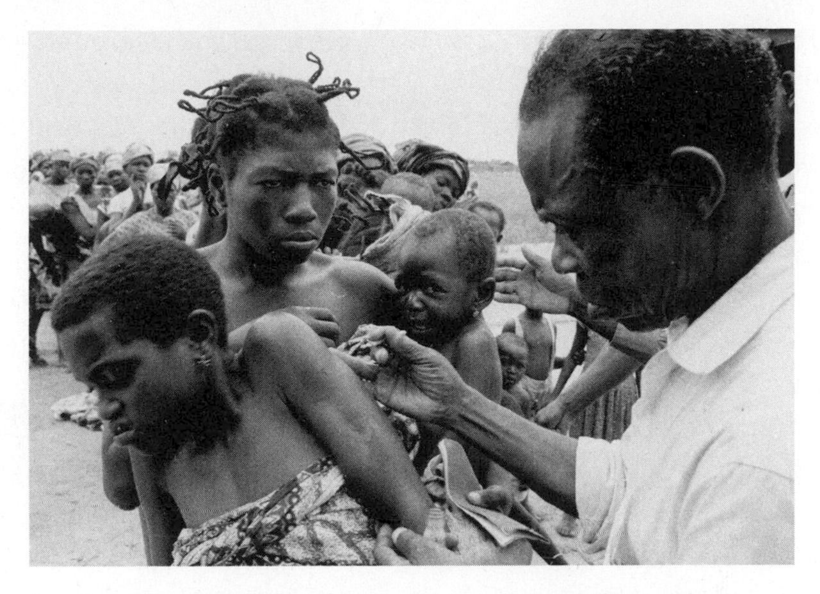

Smallpox vaccinations being administered outside Kintambo Hospital during a smallpox outbreak in Leopoldville, Republic of the Congo, 1962. (© *WHO Archives*)

could theoretically transmit that virus to many other individuals in just this way, the conditions under which Marx and his colleagues think that the global launch of HIV began in earnest.

It's important to note that Marx's work is distinct from the oral polio vaccine hypothesis for HIV origins that appeared first in a *Rolling Stone* article in 1992. Marx and his colleagues suggest that unsafe injection practices helped to spread HIV; they do not, however, argue that these techniques contributed to its introduction from chimpanzees to humans. Alternately, the OPV hypothesis argued that since oral polio vaccine was grown on fresh primate tissues HIV jumped directly from such tissues to vaccines and spread as they were administered.

The OPV hypothesis is no longer taken seriously by the scientific community for four primary reasons: (1) retrospective analysis of the original vaccine stocks showed no evidence they were infected by the chimpanzee virus that seeded human

HIV; (2) genetic analyses suggest HIV has been around for roughly one hundred years, far predating the period of OPV use; (3) the chimpanzee strains in the region where the purportedly contaminated vaccine stocks were produced are distinct from the chimpanzee virus that seeded HIV; (4) the pervasive human exposure to these viruses through hunting and butchering of wild primates provides a more parsimonious explanation for the distribution of the multiple primate viruses in the HIV family that have crossed into humans.

Further highlighting the end to the OPV debate in 2001, four separate articles published in the leading scientific journals *Nature* and *Science* laid it to rest. Doing so was important for a number of reasons, including the fact that the hypothesis was misinterpreted in a way that severely compromised ongoing vaccine campaigns, which use vaccines that are universally acknowledged as both safe and effective. An accompanying editorial to the *Nature* articles sums it up well: "The new data may not convince the hardened conspiracy theorist who thinks that contamination of OPV by chimpanzee virus was subsequently and deliberately covered up. But those of us who were formerly willing to give some credence to the OPV hypothesis will now consider that the matter has been laid to rest." More blunt were the words of my colleague Eddie Holmes, one of the world's most distinguished virologists, who said, "[The] evidence was always flimsy, and now it's untenable. It's time to move on."

While unsafe injections may very well have contributed to HIV's spread, as they did with HCV, they did not lead to its introduction. This, however, doesn't mean that we should ignore vaccine safety.

Right now around one in fifteen individuals reading this book are infected with a virus that jumped into them from a monkey. To be more specific, if you're one of these individuals, you're infected with SV40, a virus of Asian macaques, and you acquired it from a contaminated vaccine.

During the 1950s and 1960s, poliovirus vaccines were grown on cell cultures derived from macaque kidneys. Some of these kidneys were infected with SV40, which then contaminated the vaccines. The results were dramatic.

In the United States alone, up to 30 percent of the poliovirus vaccines in 1960 were contaminated with the virus. From 1955 through 1963, roughly 90 percent of American children and 60 percent of American adults were potentially exposed to SV40—an estimated ninety-eight million people. And the virus is not a trivial one. It causes cancer in rodents and can make human cells living in laboratory cell culture reproduce abnormally, a worrying sign that they may have the potential to cause cancer.

The idea that more than half of the American population was placed at risk from infection with a novel monkey virus had a notable effect on science, and epidemiologists scrambled to determine if the individuals who'd received the virus had cancer. Fortunately, while the evidence is still debated to this day, it appears clear that SV40 did not pose a serious risk for cancer and, perhaps even more importantly, it didn't have the potential to spread. We dodged a major bullet.

But because a single vaccine stock can be administered to thousands of individuals, we must remain vigilant. Contamination of a vaccine stock or multiple vaccine stocks can contribute to millions of infections with new viruses, just as we witnessed in the 1950s and 1960s with SV40. This does not mean that vaccines are not safe. They are! And they are essential to protect billions of people all over the globe. Health monitoring and vigilance in vaccine production has also never been greater. In a recent and important study, my collaborator Eric Delwart, a San Francisco–based scientist who perfects techniques to discover unknown viruses, showed that some of these new approaches, which we'll discuss in chapter 10, can be applied to even further increase the safety of vaccines. The risks associated with current vaccines are substantively less than the risks associated with the diseases they protect against. Yet the risks are not zero. We must make sure that when we knowingly connect animal and human tissues—particularly on an industrial scale—we do so with the utmost care.

Since the 1920s when Voronoff conducted his monkey gland operations, our planet has witnessed an explosion in the use of transfusions, transplants, and injections. These wonderful technologies have helped rid us of some of our deadliest diseases. Yet they have also provided powerful new biological connections between individuals, which sometimes serve as an unwelcome by-product of these beneficial tools. They provide bridges on which microbes can move, bridges that did not exist until now. They pull humans together into a completely new kind of intimate species, one unique to life on our planet and one that fundamentally changes our relationship with the microbes in our world.

➤ 8 ➤

VIRAL RUSH

Imagine the following. In a large city, let's say Manila, residents in a densely packed residential district report foul odors to local environmental health offices. Some hours later small pets begin to fall sick. Veterinarians confirm an uptick in the number of sick animals in the neighborhood. Around twenty-four hours after the first calls reporting the strange smell, local physicians note an increase in patients reporting blisters and ulcers on their skin. A few individuals report nausea and vomiting.

At around forty-eight hours the first patients hit the emergency rooms. They have fever, headache, shortness of breath, and chest pain. Some of them appear on the verge of going into shock. At the same time some of the individuals with nausea are getting worse—they're experiencing bloody diarrhea.

As the days go on the numbers increase. By the end of the first week, nearly ten thousand individuals have been hospitalized. Over five thousand of these people have died painful deaths. At the end they can barely breathe—their skin blue from lack of oxygen. Eventually, septic shock and severe brain inflammation strike, killing most of them. As the number of deaths increase, journalists flock to the scene. Manila residents attempt a mass exit, and despite the best intentions of the government, the city verges on the brink of widespread and crippling panic.

The case I've outlined is a hypothetical one. But just barely. In June 1993 the Aum Shinrikyo cult aerosolized a liquid suspension of *Bacillus anthracis* from the top of an eight-story building in the Kameido neighborhood in the eastern part of Tokyo. They launched a bioterror assault on one of the largest and most densely packed cities in the world.

The good news is that they failed. An analysis written in 2004 states that their choice of a relatively benign anthrax strain, low concentrations of the bacterial spores, an ineffective dispersal system, and a clogged spray device all served to make the 1993 incident in Tokyo a flash in the pan. No humans got sick, although some pets appear to have died as a result of the release.

If the Aum Shinrikyo cult had come across a more deadly version of anthrax and used even slightly better dispersal systems, things could have turned out quite similar to our hypo-

Asahara Shoko, the founder of Aum Shinrikyo, praying with followers in India. (© *Georges de Keerle / Sygma / Corbis*)

thetical scenario above. We know that the apocalyptic cult had looked for more than just anthrax. The group set up multiple laboratories and dabbled in cultivating many agents. They played with botulinum toxin, anthrax, cholera, and Q fever. In 1993 they led a large group of doctors and nurses to the Democratic Republic of Congo, ostensibly on a medical mission, but actually in an attempt to bring back an isolate of the Ebola virus for use in their grim operations.

Even if they had succeeded in their anthrax release, the deaths and disruption caused by Aum Shinrikyo would have been restricted to the individuals exposed to the spores they released. Anthrax does not transmit from person to person. Though deadly, it is not contagious. But anthrax is only one of many agents that could be used by terrorist groups. Bioterror is among the most serious concerns for security experts. It is an ideal tool for the weaker parties in so-called asymmetrical warfare, where enemies differ significantly in the resources and firepower they can draw on for battle. Even a weak opponent, like a terrorist group, can wreak havoc with the right combination of microbe and dispersal.

Microbes hold great potential for terror groups. They are much easier to gain access to than chemical or nuclear weapons. And, critically, unlike either chemical or nuclear weapons, they can spread on their own. They can go *viral*, which is something that neither deadly sarin gas nor a dirty bomb could accomplish. Perhaps the only comparable situation is the long-term horror of some nuclear fallout expressing itself in generations of mutated offspring and high rates of cancer, as seen in Hiroshima. But those insidious effects are environmental and thus relatively slow. A fast-acting, fast-spreading viral weapon would have that impact over days, not decades.

It would be a mistake to underestimate the risk for bioterror, and most who study it contend that it is just a matter of time before it's unleashed on a human population.

The fact that deadly microbes can be made to proliferate under lab conditions, whether in legitimate laboratories or fly-by-night terrorist workshops, adds another dimension to global pandemic risk. While exceptionally unlikely, if terrorists ever got their hands on one of the few remaining vials of smallpox, the results would be devastating. While smallpox has been eradicated in nature, two sets of smallpox stocks remain under lock and key—one at the U.S. Centers for Disease Control in Atlanta and one in the State Research Center of Virology and Biotechnology (VECTOR) in Koltsovo, Russia. Both facilities have high containment bio-safety level 4 facilities. There's been debate about possibly destroying the remaining stocks in these labs, but to date the decision has been deferred because of the potential benefit of access to live virus for the production of vaccines and drugs.

Interestingly, in 2004 scabs from suspected smallpox were found in Santa Fe, New Mexico, in an envelope labeled as containing scabs from vaccination. The finding points to the possibility of other unknown lots of smallpox existing in a lab freezer or somewhere else. If they were released purposely or accidentally, the consequences would be devastating. Since smallpox has been eradicated, we no longer inoculate against it. So, for smallpox, such a release would be a perfect storm. For us, it would be catastrophe.

Another risk is what is increasingly referred to as "bioerror." Unlike bioterror, bioerror occurs when an agent is released accidentally but spreads widely. In 2009 Don Burke, the mentor of my postdoctoral fellowship, published a paper on the emergence of influenza viruses. In it he analyzes a variety of influenza viruses that have spread in humans. One of the more interesting examples is the November 1977 epidemic that affected the Soviet Union, Hong Kong, and northeastern China. The virus involved was nearly identical to a virus from an outbreak over twenty years before, and it hadn't been seen since. Don and his colleagues echoed earlier research on the virus noting that the most likely explanation was that a lab strain had been accidently reintroduced into the lab workers and had spread from there.

Over the coming decades, as it becomes possible for the masses to have access to detailed biological information and the techniques to make or grow simple microbes, the probability of bioterror and bioerror will only grow. While most people normally think of biology as occurring primarily in secure labs, this may not always be the way it works. In 2008 two teenage girls from New York City sent away specimens of sushi to the Barcode of Life Database project, a fascinating early program to simplify and standardize genetic testing. They wanted to determine if the high-priced fish that they were buying was what it was sold as. They found that often it wasn't. But they also found a way to get genetic information that until then was only available to scientists.

But the student sushi study was about more than discovering that some of the sushi vendors in New York City rip off their clients. It was one of the first notable examples of nonscientists "reading" genetic information. Early in the information technology revolution, only computer programmers could read and write code, like HTML. Then nonprogrammers began to read code, then write code, and now we all regularly read and write code on blogs, wikis, and games. As with any system of sharing information, what starts as something highly specialized often becomes universal. In the not-too-distant future, the small group of people conducting do-it-yourself biology may become the norm. In that world the need for monitoring to control bioerror will be more than just theoretical. In a famous prediction made by Sir Martin Rees, the former president of the Royal Society of London warned, ". . . by the year 2020 an instance of bioerror or bioterror will have killed a million people." The chemistry to create a pipe bomb or a meth lab becomes the biology to create a viral bomb.

In this chapter we will explore the next big killers—the microbial threats that keep me awake at night. Certainly, bioerror and bioterror are among them. The frequency of both of these threats will rise in the coming years, but at least for the moment, the greatest risks we face are still those that exist in nature.

In some biological arenas, the age of discovery is over. We know the rate at which we'll discover new species of primates, for example, will be very low indeed. For viruses, that's not the case. My collaborator Mark Woolhouse, one of the early leaders in the field of emerging infectious diseases, has put together real numbers on this. He and his colleagues have plotted the rate of discovery of new viruses since 1901. Their analysis suggests we're nowhere near the end of viral discovery; we'll find on average one or two viruses per year over the next ten years, and that's likely a conservative estimate.

One of the reasons contemporary scientists are finding new viruses is that we're looking. Studies like the ones conducted by my research group, which we'll discuss in the coming chapters, actively seek to find unknown viruses in humans and new viruses lurking in animals that might be the next to jump. Genetic techniques for uncovering the unknown microbial world are also advancing, which makes finding these new agents easier and faster than ever. But intensive research and heightened attention are not the only reasons we're seeing new things.

The combination of factors we've discussed in the previous chapters has created the perfect conditions for maintaining new agents in the human species. We live in a massively interconnected world. Links made by transport networks and medical technologies radically increase the probability that an animal virus that enters into us—no matter where—will be able to gain a foothold and spread. This means that while some of the new things we're finding might have crossed over in the past, they haven't persisted. From our perspective, they're new.

On February 21, 2003, a man at the Metropole Hotel in Hong Kong was sick—very sick. He had come from the nearby Guangdong province and had arrived at the upscale hotel, which has a fitness center, restaurants, a bar, and a swimming pool. He stayed just one night in the now infamous room 911. And he would become among the most famous "super-spreaders" of modern history.

A *super-spreader* is a person (or animal) who plays an outsized role in the spread of an infectious disease. The resident of room 911 at the Metropole had severe acute respiratory syndrome, or SARS, and his virus spread to at least sixteen other individuals. They in turn spread the virus to hundreds of other individuals as they dispersed to the far points of the globe—Europe, Asia, North America. Even three months later, investigators were able to pull genetic information of the virus from the carpet near room 911, information that likely got there from his coughing, sneezing, or vomiting.

We do not know exactly how the resident of room 911 became infected with the SARS virus. It may have been through contact with an infected animal. We now know that SARS ultimately originated in bats. Because people in the Guangdong province commonly eat wild animals and purchase them in live animal markets, or *wet markets*, the resident of room 911 may have had contact with an infected bat purchased in one such market. Alternatively, he may have acquired the virus from a civet, a small carnivore and a delicacy in that region of China. By that time, civets had acquired the SARS virus from bats. Or he may have been infected from a person who had acquired the animal virus. Perhaps most likely the virus had spread undetected for some time before he got it himself.

However the Metropole guest acquired the virus, his illness appears to have sparked the SARS pandemic that would follow, a pandemic that would go on to infect thousands of people in at least thirty-two countries on every inhabited continent and have an economic impact measured in billions of dollars. The SARS pandemic provides a perfect example of how our modern world cultivates pandemics.

Hong Kong has a higher density of people living in it than almost any other city in the world and certainly higher than any city that existed prior to the twentieth century. Thousands of international flights going to just about any part of the world you can imagine originate in Hong Kong every day. It also sits a short drive from the Guangdong province of China. Guangdong houses hundreds of millions of people and its culinary history includes wild animal delicacies and dishes like pig organ soup.

The combination of high human population densities, intense livestock production, close contact with the diverse microbes of wild animals, and a massive, efficient transportation network gives us a good sense of where the world is heading with regard to pandemics. Hunters begin the process by capturing wild animals and bringing them to markets, some of which exist in highly urban areas. The wet markets, which house live animals, pose particular risks. Once an animal has been killed, the microbes within it also begin to die, but if a living wild animal makes it to one of these urban markets, the entire panoply of its microbes are placed squarely in the midst of large numbers of humans. A virus that gets out here has definitely won the microbial lottery.

While an interesting example, Guangdong is by no means unique. Regions that house important wildlife diversity are urbanizing at rapid rates throughout the entire world. Within the past few years, for the first time in human history, we became a primarily urban species—more than 50 percent of the human population now lives in urban areas, and that number is growing. By 2050 it has been estimated that 70 percent of the world's population will live in cities. And when highly dense urban populations, the microbes of wild animal and livestock populations, and efficient transportation networks overlap, new diseases will inevitably emerge.

In Africa the particular course of development has provided another set of unique microbial risks. In central Africa, a region where I lived and worked for a number of years, the combination of urbanization, deforestation, road building, and consumption of wild game are conspiring to create a recipe for disease emergence.

One of the most common economic activities among the Congo Basin countries is logging. Unlike the clear-cutting that characterizes logging in some parts of the world, in central Africa most logging is selective. In selective logging, roads are cut into the relatively pristine regions with valuable trees, and workers are transported into them to extract the timber.

Logging in this way has a number of consequences for how viruses emerge. Among the first things that occur when a new logging camp opens is the large influx of workers. People arrive to clear roads, cut tracks, fell trees, haul trees, cut them, load them, and manage camps; they all come together to make temporary towns. The towns consume meat, and since most of the meat consumed in rural forested regions of central Africa is from wild game, local demand for hunting increases. This attracts more hunters and incentivizes them to hunt more. All of this serves to increase the number of animals caught and, therefore, the human contact with the blood, body fluids, and corresponding microbes of the animals present in these biodiverse habitats.

Logging trucks in southern Cameroon. (*Adria Prosser*)

The existence of logging roads also leads to fundamental changes in the way that people can hunt. Historically, hunters lived in villages. Their daily hunting would radiate in a circular fashion from these villages, with decreased impact at the periphery of the hunting range. Logging roads provide a greater number of points at which hunters can enter the forest, lay traps, and make kills using firearms. This has been demonstrated through detailed studies in and around the Campo Ma'an National Park by the Cameroonian ecologist Germain Ngandjui. At the same time that forest access is increasing, the movement of trucks along the roads provides increased routes to urban markets, which in turn increases the number of hunters who engage in the practice.

Whether from the pressures of the workers themselves or the roads they create, the practice of logging changes the frequency at which humans have contact with wild game. The more contact that occurs, the better the chance that a new agent will jump over. This is compounded by the interconnectivity discussed in chapter 6. The villages are remote, but they are connected by road to major ports, where the logs (and microbes) can be put on ships and moved throughout the world.

Our work in some of the most rural regions in central Africa provides clear evidence that even seemingly remote places are most definitely on the grid. We regularly screen for potentially pandemic viruses like influenza, and we see evidence of the globally circulating pandemic H1N1 even in villages in the middle of the forest. And while we certainly see unusual viruses that are local, we also see cosmopolitan strains of HIV that have worked their way down the road to infect people living in distant rural lands. New agents can increasingly get in and out of even the most remote locales.

Sometimes multiple factors accumulate to compound the emerging pandemic threats. This is exactly what's happened with the global spread of HIV and its associated impact on the human immune system. As we've discussed, HIV originally entered into humans from chimpanzees almost certainly through the hunting and butchering of these animals by people in central Africa. But now that it's in human populations, spreading and infecting such a large number of us, it has the potential to alter the emergence equation.

Among the terrible consequences of AIDS is that it hampers the immune system. In fact, when people die of AIDS, they don't die of HIV per se. They die because they eventually succumb to infectious diseases that their immune systems can no longer control. Approximately 1 percent of the human population worldwide is immunodeficient. While malnutrition, therapies for cancer, and organ transplantation play a role, the most significant factor is global infection with HIV.

Immunodeficiency leads to the proliferation of a whole range of usual suspects. Agents like tuberculosis and salmonella multiply more effectively in immunosuppressed people. Common agents that aren't normally deadly can become fatal when immune systems are weak. Viruses like cytomegalovirus and human herpesvirus 8 afflict AIDS sufferers. But immunosuppression can also provide an entryway for new agents.

Most animal agents don't come preadapted to humans. Even microbes from some of our closest relatives often require a combination of genetic changes in order to be able to survive and spread in a human host. So when a highly exposed person like a hunter contracts a new agent, the infection will generally be fleeting. Yet in an immunocompromised host, quickly evolving microbes can often gain precious time, free of immune pressure, to go through a few more generations of reproduction, increasing the probability that they will come upon the right suite of adaptations necessary to take hold in a new species.

And it doesn't stop there. Sometimes a new virus will cross over into someone who has been exposed to an animal, but the virus will go nowhere. The existence of numerous immunosuppressed people in a community will, however, increase the chance that the virus can begin the process of spreading once it adapts to humans. Immunosuppression, as caused by HIV or another compromising agent, provides another foothold for new microbes as they cross the elusive species barrier.

This risk is not trivial. In 2007, along with my colleagues, I reported the results of a study we'd done in Cameroon to determine the rate of HIV in individuals who had contact with wild animals through hunting or butchering. We analyzed data from 191 HIV-infected people living in rural villages near forested settings. The vast majority of the individuals we studied reported butchering and consuming wild animals. Over half of the people reported butchering monkeys or apes. Most worrying, 17 of the HIV-positive individuals reported injuries while they'd hunted and butchered wild animals—perfect opportunities for direct blood-to-blood contact and bridging of blood-borne microbes.

The fact that people in direct contact with the blood and body fluids of wild animals also have HIV and may be immunocompromised represents a serious risk for the emergence of new microbes. Hunting and butchering provide opportunities for contact with the microbes present in virtually every animal tissue. When these agents are regularly in contact with people with limited defenses, it may provide a shortcut for microbes as they traverse the boundaries between species.

•

Hunting and butchering create serious risks, but even contemporary industrial livestock practices, including factory farms and modern meat production, substantially alter the ways in which we interact with animals in our world. They also increase the probability that an animal virus will spill over into humans and become a pandemic.

Livestock production has changed dramatically over the past forty or so years. One of the major changes has been raw numbers (so to speak). There are now more than one billion cattle, one billion pigs, and over twenty billion chickens living on our planet. There are estimated to be more domestic animals alive today than in all the past ten thousand years of domestication through 1960 combined. Yet this is not simply a numbers game. How the animals are grown and grouped has also dramatically shifted.

In 1967 the United States had around a million pig farms. As of 2005, the number had shrunk to a little over one hundred thousand. More pigs and fewer farms means that more and more pigs are packed together on single large-scale industrial farms. The same trends exist with other livestock species. In the United States four massive companies produce over half of the cattle, pigs, and chickens. And this is not limited to the United States. More than half of the livestock produced globally now originate in industrial farm settings.

While it's more economically efficient to grow livestock in industrial settings there are consequences for microbes. As we've seen with humans, larger numbers of livestock grouped more closely together increases the capacity of livestock populations to maintain novel microbes. The animals living on massive industrial farms largely do not exist in a state of perfect isolation. Contact with blood-feeding insects, rodents, birds, and bats all provide the opportunity for new agents to enter into these incredibly massive colonies of animals. When they do, the industrial farms become far more than settings to grow meat. They become incubators for infectious agents that could move into human populations. We have seen this occur with Nipah virus in Malaysian pigs, as discussed in chapter 4. Other viruses like Japanese encephalitis and influenza can act in similar ways.*

The number of livestock on the planet now boggles the mind, but the way that they're transformed into meat also differs in important ways from how it's been done since domestication began. Historically, a single animal would feed a family or at most a village. With the advent of processed meats, a single hot

* While the changes we've experienced with large-scale industrialization of livestock production currently outweigh the benefits when it comes to microbes, that is not a necessary outcome. Industrial scale efficiencies in animal farming have the potential for better disease monitoring, and if done well could ultimately ensure that domestic animals remain separated from wild animals. Industrialization also serves to decrease the number of humans who have contact with living animals, which decreases the points when microbes can spill over. At the far end of this continuum would be fully artificial, or in vitro, meat. In vitro meat is animal flesh grown on cultures entirely independently from animals. The idea of cultured meat is currently unappetizing to many, and the health and other risks must be examined in depth. Nevertheless, it could have amazing benefits. Cheap factory-produced in vitro meat could address hunger issues and decrease the need for use of domestic and wild animals for meat, a situation that would radically decrease the introduction of novel microbes. Decreasing contact with domestic animals means decreasing contact with their microbes and the microbes they've acquired from their wild kin.

dog consumed at a baseball game can consist of multiple species (pig, turkey, cattle) and contain meat derived from hundreds of animals. When you bite into that hot dog, you're literally biting into what was only a few decades ago an entire farm.

Combining the meat of many animals and then distributing it to many people has obvious consequences. Connecting thousands of animals with thousands of consumers means that an average meat eater today will consume bits of millions of animals during their lifetimes. What previously was a direct connection between one animal and one consumer is now a massively interconnected network of animal parts and those that eat them. And while cooking the meat certainly eliminates many of the risks, the massive number of interactions increases the potential that a rogue agent will make the jump.

This is what appears to have happened in the case of the sheep disease scrapie and bovine spongiform encephalopathy (BSE), better known as mad cow disease. BSE is among the fascinating group of infectious agents known as prions, mentioned in chapter 1. Unlike viruses, bacteria, parasites, and any other group of life we know of on the planet, prions lack the genetic blueprints of biology (i.e., RNA and DNA). Rather than the combination of genetic material and proteins that make up all other known life, prions simply have protein. While this may seem insufficient to accomplish any organic task, prions are capable of spreading. And they can cause serious disease.

BSE was first identified as a novel cattle disease in November 1986 because of the dramatic symptoms it causes in cows. They walk and stand abnormally, and after some months they experience violent convulsions and death. While there's still some debate about its origins in cows, it appears that it came from sheep. During the 1960s and 1970s as the development of cattle feed was industrialized, one type of cow feed involved the rendering of sheep carcasses into meat and bone meal. Sheep have long been known to have a prion disease called scrapie, and it appears that processing their carcasses as cattle feed permitted the agent to jump over and adapt.

Once it jumped to cattle, BSE then spread through more feed. Some cattle carcasses, like sheep carcasses, are also ground into feed for cattle. It appears that once the prion crossed from sheep to cattle, its primary communication was through infected cattle meat and bone meal processed for the next generation of cows.* The spread was remarkably effective. Some have suggested that during this period more than a million infected cows may have entered into the food chain. But not all of these prions stayed in cows.

Around ten years after the first identification of BSE, physicians in the UK began to recognize a fatal neurodegenerative disease among humans who were potentially exposed to contaminated beef. The patients showed evidence of dementia, severe twitching, and an increasing deterioration of muscle coordination. Evidence from the patients' brains revealed that they had been ravaged in exactly the same ways as those of the cows. Experimental evidence showed that the disease could also be transmitted to primates whose brains were inoculated with brain tissue from infected humans. These human patients had been infected with BSE, but when found in humans, the same disease is called variant Creutzfeldt-Jakob (vCJD) disease.

While only twenty-four human cases of vCJD have been

---

* Cows are not the only species that have acquired prions by consuming their own kind. Another fascinating prion, kuru, a fatal neurodegenerative disease, moved in exactly this way among the Fore people of the Eastern Highlands Province of Papua New Guinea. The Fore practiced ritual cannibalism, consuming relatives and community members who had died and smearing the deceased's brains on their bodies to help free their spirits. These mortuary feasts were determined to be the way kuru was transmitted. Following the prohibition of ritual cannibalism in the 1950s, the epidemic has now effectively come to an end.

confirmed to date, there are certainly others, as the definitive diagnosis is difficult to make. Much is still unknown about vCJD, but it's increasingly suspected that infected humans must have both genetic susceptibility for the deadly brain disorder as well as exposure to infected cow tissue. Analysis of the tonsils and appendixes removed from healthy patients suggests that as many as one in four thousand people who were exposed during the UK BSE epidemic are carriers who show no sign of disease. This is particularly worrying since vCJD has been shown to pass through organ transplantation and may also pass through blood transfusions.

The way that we now grow and distribute meat differs fundamentally from how we did it in the past. We also transport live animals in new ways. The relative ease of international shipping means that people can move livestock from regions that were once remote. And the situation is not unique to animals. Many of our plant food sources are now transported thousands of miles and eaten by millions before any microbial contamination related illness would be detected.

In chapter 6 we discussed how monkeypox rates are rising in DRC. But monkeypox has not been restricted to Africa. In 2003 monkeypox hit the United States. Careful investigation of the 2003 US outbreak showed that it emerged from a single pet store—Phil's Pocket Pets of Villa Park, Illinois. On April 9 of that year, around eight hundred rodents representing nine different species were shipped from Ghana to Texas. The shipment included six different groups of African rodents, including Gambian giant rats, brush-tailed porcupines, and multiple species of mice and squirrel. Subsequent testing by the CDC showed that Gambian giant rats, dormice, and rope squirrels from the shipment were all infected with monkeypox, which likely spread among the animals during shipment. Some of the infected

Gambian rats ended up in close proximity to prairie dogs at the Illinois pet store, and those prairie dogs appear to have seeded the human outbreak.

Over the following months there were a total of ninety-three human cases of monkeypox in six midwestern states and New Jersey. And while most of them probably resulted from direct contact with infected prairie dogs, some may very well have resulted from human-to-human transmission.

The moving and mingling of animals as pets and food increases the probability that new agents will enter into the human population. It also increases the chances that distinct microbes will end up in the same host and exchange genes. As discussed earlier, there are multiple ways in which a virus can change genetically: direct changes in genetic information (mutation) or the exchange of genetic information (recombination and reassortment). The first option, genetic mutation, provides an important mechanism for slow and steady production of genetic novelty. The second options, genetic recombination and reassortment, provide viruses with the capacity to quickly gain entirely novel genetic identities. When two viruses infect the same host, they have the potential to recombine, exchanging genetic information and possibly creating a completely new "mosaic" agent.

This has already occurred to important effect. As we learned in chapter 2, HIV itself represents a mosaic virus—two monkey viruses, which at some point infected a single chimpanzee, recombined and became the ancestral form of HIV. Similarly, influenza viruses have the capacity to pick up entirely new groups of genes by forming these mosaics through reassortment, where entire genes are swapped.

Influenza viruses can reassort on the farms where humans, pigs, and birds interact. Pigs have the potential to acquire some human influenza viruses. They also can acquire viruses from birds, including wild birds that may pass through on migration routes. These wild birds can infect pigs directly or indirectly through domestic birds such as chickens and ducks. When new viruses from birds interact with human viruses in an animal such as a pig, one of the outcomes is a completely new influenza virus with some parts from the circulating human virus and some parts from the bird virus. These new viruses can spread dramatically when reintroduced into human beings since they can differ sufficiently to avoid detection by natural antibodies and vaccines from earlier circulating influenza strains.

Recombination plays a potentially vital role in a number of viruses. Genetic analyses of SARS show that it's likely a recombinant virus between a bat coronavirus and another virus, probably a separate bat virus we have yet to discover. These two viruses formed a novel recombinant mosaic virus prior to infecting humans and civets. These viruses' potential to recombine may very well have related to the interaction of animals that previously would never have been in contact in the wild, as they made their way along market networks.

My mentor Don Burke, who now leads the University of Pittsburgh's School of Public Health, has played a pivotal role in pointing out how recombination between viruses can help seed new epidemics. He coined the term *emerging genes* to refer to this process. Historically, virologists thought that new epidemics result from the movement of an entire microbe from an animal to a human. As we've seen in HIV, influenza, and SARS, recombination and reassortment provide other more stealthy methods to seed new epidemics. Rather than transplant an entire new microbe, two microbes, one old and one new, can temporarily interact in a single host and exchange genetic material. The resulting modified agent may have the potential to spread and become a completely new, and completely unprepared for, pandemic. In these cases it's actually newly swapped genetic infomation that causes the pandemic rather than a new microbe—hence the term *emerging genes.*

In the coming years we'll see more and more pandemic threats. New infectious agents will spread and cause disease. New pandemics will emerge as we go deeper into the rain forests and unleash the agents previously unconnected to international transportation networks. These agents will spread as dense population centers, local culinary practices, and wild-animal trade increasingly intersect. The impact of epidemics will be augmented by HIV-caused immunosuppression that increases the risk of new agents adapting to a damaged human species. As we move animals quickly and efficiently around the world, they will, in turn, seed new epidemics. Microbes that have never encountered each other now will, and they'll form new mosaic agents capable of spreading in ways that neither of their parents could manage. In short, we'll experience a wave of new epidemics, ones that will devastate us if we don't learn to better anticipate and control them.

PART III

THE FORECAST

► 9 ◄

VIRUS HUNTERS

On December 9, 2004, primatologists working in the Dja Biosphere Reserve in southern Cameroon collected specimens from a dead chimpanzee. The chimp was sprawled out on the forest floor, eyes closed, but seemingly unmolested by a human or other predator. The team was rightfully concerned.

Belgian scientist Isra Deblauwe and her Cameroonian colleagues had started their long, tedious work some three years earlier. In the tradition of primatologists like Jane Goodall, their goal was to study wild great apes, our closest living relatives, to learn about them and ourselves.

And a few years later, they produced some interesting results. The team reported that, like other chimpanzee populations, the chimpanzees in the Dja used tools. In particular, they modified sticks to extract honey from underground bee nests. Chimpanzees, like all apes, including ourselves, love honey, and the information from the Dja team would add to the understanding of how different chimpanzee cultures use tools in different ways.

But on that rainy December day in 2004, honey was the last thing on the scientists' minds. Four days after taking samples from the first dead chimpanzee, they took samples from another. Then on December 19 they took samples from a dead gorilla. This was worrying. Since the primatologists only followed a fraction of the population of apes in the Dja, what they'd seen was likely to be just the beginning. Many other unidentified apes might be dead, valuable wild kin whom the team had spent years working to understand. The consequences for conservation and research could be considerable.

But the threat to wild apes, while significant, was not the only problem. The researchers knew that the Ebola virus had wiped out large numbers of apes in Gabon, only a few hundred kilometers to the south. Ebola not only kills chimpanzees but from time to time has also jumped to humans causing dramatic and potentially epidemic-inducing cases. They also knew that one of their primatologist colleagues had acquired Ebola in the Ivory Coast when investigating deaths just like these. Whatever caused these ape deaths was not to be taken lightly.

Fortunately, they had responded according to a plan. First and foremost, the primatologists knew that they should not directly touch the carcasses. Months earlier, when the first dead animals had been seen, they had sent a message to colleagues in Yaoundé, Cameroon's capital. The message in turn was transmitted to Mat LeBreton, the dedicated and skilled biologist who leads our ecology team and has pioneered a number of new techniques in viral ecology. Based in Yaoundé, LeBreton helped support an international team, including relevant ministries and laboratories in central Africa and Germany on the outbreak investigation that would follow.

The investigating team rapidly put together and deployed a mission to the Dja, a stunningly beautiful and unique rain forest habitat located along one of the major tributaries of the massive Congo River. There they worked with the primatologists to collect the specimens. They managed to obtain specimens from the skull and shoulder of the first chimpanzee. They also collected a specimen from the leg of the second chimpanzee, the jaw of the gorilla, and some muscle from a fourth victim—a chimpanzee—who died in early January 2005.

The safely preserved specimens then made the trip to expert laboratories. They went to the high containment laboratory of Eric Leroy, the virologist whom we worked with to discover the new strain of the Ebola virus discussed in chapter 5. The specimens also went to our collaborator Fabian Leendertz, a veterinarian and microbiologist working at the Robert Koch Institute in Germany who has perfected the study of ape microbes during many years spent shuttling between field sites in Africa and his lab in Berlin.

The results were surprising. While we all had come to assume that the same wave of Ebola knocking down ape populations south of the border in Gabon had killed the animals in the Dja, the specimens all came back negative for the Ebola virus. They were, however, all positive for another deadly agent—anthrax.

In 2004 Leendertz and his colleagues had reported a similar die-off of chimpanzees in the Taï forest of the Ivory Coast due to anthrax. So while the gorilla death in the Dja was the first of its kind, anthrax was already known to be a killer of forest apes. Strange perhaps but not unprecedented. How exactly a bacteria normally found in grasslands ruminants got to the apes in the Dja and Taï forests is still a mystery. There were some theories. Anthrax spores remain viable for long periods of time, even up to a hundred years. The spores can contaminate water supplies, so the apes may have picked it up from lakes or creeks. They may also have become infected while hunting or scavenging on ruminants, such as forest antelopes, that had themselves been infected. Or perhaps, at least in the Taï forest outbreak, neighboring farms had seeded the outbreak when the apes had foraged for food in cropland contaminated by anthrax from cattle.

Whatever the route of infection, the findings from the Dja

Gorilla killed by anthrax in the Dja Biosphere Reserve, Cameroon. (*Matthew LeBreton*)

and from the earlier animal epidemic in the Ivory Coast showed that the declining populations of African apes had more than hunting and habitat loss to blame. Viruses like Ebola have swept through large swaths of the remaining habitats of wild apes, and now anthrax must also be considered a threat to these valuable wild animals. From a personal perspective, having worked with wild chimpanzees and helped habituate populations of gorillas in Uganda, I feel that the mounting threats to our closest living relatives is a tragic loss to the heritage of our particular form of life.

From the perspective of my work tracking and preventing pandemics, the deaths pointed out another glaring weakness in the way that we catch these epidemics. The discovery of anthrax in the Dja forest did not represent a success in pandemic prevention. It was rather the epidemiological equivalent of dumb luck. Only a trivial fraction of the global ape populations are under the watchful eye of woefully underfunded primatologists. If we're relying on these scientists to regularly capture the animal epidemics that could signal future human plagues, then we're destined to fail. To truly catch epidemics early, we'll need much more.

How can we hunt down deadly viruses and control them? A few primatologists finding dead animals is not a surveillance system. But what is the right way to catch new epidemics and stop them before they spread? This section will explore just that: the contemporary science of pandemic prevention. It will discuss the ways that my team and other colleagues and collaborators are working to create systems that will be able to catch and stop new epidemics before you (or CNN) even know about them. Preventing pandemics is a bold idea, yet no bolder than when cardiologists in the 1960s began to think that they could prevent heart attacks, a medical advance that was radical at the time but is now largely taken for granted.

My own thinking on this dates to the late 1990s when I joined Don Burke at Johns Hopkins and agreed to establish a field site aimed at monitoring humans and animals for new viruses in central Africa. It was an exciting time, when the idea that simply responding to pandemics would no longer be sufficient was truly in the air. I remember spending many afternoons in Don's office alternating between urgent scribbling on the whiteboard and thinking out loud about what would be needed to accomplish the task.

Among the ideas that we generated during that time, one lasting concept particularly stands out: *viral chatter*. When he coined the term, Don did so as a direct parallel to *intelligence chatter*. One way of thinking about this is to ask the question: how do security services prevent terror events?

Intelligence services use a range of techniques to monitor for potentially threatening events, but among their most valuable tools is the monitoring of chatter. Intelligence agencies scanning e-mails, phone calls, and online chat rooms can follow the frequency that certain signals occur. If a journalist were to fire off an e-mail that included the words *al-Quaeda* and *bomb* for example, it would be picked up by an automated system that filters for suspicious key words. Having said that, it would not likely make it to the desk of an analyst, since the systems also register e-mail accounts and IP addresses and would hopefully flag the chatter with *journalist.*

During testimony on the September 2001 attacks on the United States, the former CIA director George Tenet said that the "system was blinking red" in the months leading up to 9/11. Similarly, although it was an accidental event, the day that the Chernobyl reactor melted down in 1986 there was a significant spike in message traffic in the former Soviet Union. Knowing what sorts of key words to look for and who the usual suspects are, as well as understanding how they communicate with each other can provide valuable intelligence to help predict rare but important events.

As Don and I considered it, we asked ourselves what a global system to monitor the viral equivalent of such chatter would look like. How could we monitor the many thousands of interactions that occur between humans and animals in order to detect the chatter events—in our case the jumping of novel viruses to humans—that would signal a looming plague?

Clearly, a system that depended on communities like the primatologists, whose primary focus was studying animal behavior and ecology, would not be sufficient. An ideal system would monitor global viral diversity in humans and animal populations and detect when agents jumped from animals to humans. While theoretically possible, such a system defied resources and technology at the time.

As we'll discuss in greater detail in chapter 10, the current laboratory methods for accurate and comprehensive surveys of the diversity of viruses in people and animals, while improving all the time, are not yet at the point of being deployed globally. Also, the simple logistics of monitoring everyone would be impossible. To begin, we'd need a much more focused system—a system focused on a small set of *sentinels*, key populations that would allow us to monitor viral chatter with the resources we currently have.

I remember vividly the first time I thought about the role of hunting in the transmission of infectious agents. While a graduate student at Harvard, I spent my first two years focused on the study of wild ape populations. Among the pleasures of being a graduate student in the Department of Biological Anthropology was being able to interact with one of the leading professors, Irv

Dr. Irv DeVore during pioneering work to study wild baboons in Kenya. (*Nancy DeVore*)

DeVore. Irv, a leading teacher and thinker in primatology and human evolution, has a striking head of white hair and a Vandyke beard to match. The son of a Texan Baptist minister, he taught human evolution with the fervor of an evangelist and is beloved among the scores of prominent scientists who benefitted from his tutelage.

From 1993 through 1995 I worked for Irv as a teaching fellow for the class he co-taught at the time with the Harvard psychologist Marc Hauser. The course, Human Behavioral Biology, was referred to as "Sex" by the Harvard undergraduates because of the focus on human reproduction. During those years, I had the opportunity to meet with Irv in his office on the top floor of the Peabody Museum and on occasion at the wonderful evolution-soaked beer hours that proliferated at faculty homes.

On one particularly memorable afternoon, I remember speaking to Irv in his wood-paneled office in the Peabody. During our freewheeling conversation, the topic reverted to my growing obsession at the time—microbes. It was then that Irv told me a story that would help put me on the research track I've taken for the last fifteen years.

During one of his summers spent on Martha's Vineyard, Irv had come across a dead rabbit while driving home. Assuming it was a healthy animal that had been killed by a car and being a lifelong hunter who had worked with indigenous hunters throughout the world, Irv did what seemed natural for him. He picked up the rabbit and brought it home, where he subsequently dressed and cooked it for supper.

Within a few days Irv was very ill. He experienced fever, diminished appetite, and severe exhaustion. His lymph nodes enlarged. Fortunately, he went to an emergency room immediately, because as it turned out he'd acquired tularemia, a potentially fatal bacteria that often infects wild rabbits and other rodents. Death occurs in less than 1 percent of people with prompt treatment, but had he not been treated quickly, he may very well have died a painful death from multiple organ failure.

Irv likely acquired tularemia when skinning the infected rabbit. A common route of entry for this bug occurs during butchering, when the bacteria can be inhaled into the lungs. By the time Irv finished his story, my mind was racing with the possibilities. One of Irv's earlier works was a book called *Man the Hunter*, and he'd spent many years living with hunter-gatherer populations in Africa, populations that don't practice farming and live exclusively on wild foods. Our conversation veered to the idea of working with these populations, who no doubt had extraordinarily high rates of exposure to the microbes in the animals around them.

In 1998, a few years after my conversation with Irv, I wrote about the role of hunting in disease transmission. In the article, I proposed that hunters could act as sentinels—and if we studied them over time we could get a sense of what, how, and when microbes were jumping into humans. During my conversations with Don Burke a few years later, this became a common point of discussion for us as we explored the concept of viral chatter. How might hunters lead us to the critical microbes making that fateful jump into the human species?

When Don recruited me to join his growing program at Johns Hopkins, he had already established a close collaboration with a Cameroonian scientist examining retroviruses, like HIV, in the region of central Africa where they originally emerged. I would spend many years working with Don and the Cameroonian colleague, Colonel Mpoudi Ngole. During those years, we would put the foundation in place for the first real system attempting to catch novel pandemics before they emerge.

One of the first people I met when I arrived in central Africa was the aforementioned Colonel Mpoudi (pronounced m-POODY),

Colonel Mpoudi Ngole.
(*Nathan Wolfe*)

a large, imposing, mustachioed man who so consistently wore a uniform that I wondered at times if he slept in it. The Colonel, as I refer to him to this day, is a quiet but incredibly productive physician and scientist. He is known by many of the people in Cameroon as Colonel SIDA (SIDA is the French acronym for AIDS) for his years of relentless work to stem the tide of the AIDS pandemic in central Africa. The Colonel has a subtle yet commanding presence, and he's used to getting his way. During my first years in Cameroon, we did battle from time to time, fighting over the best way to use scarce resources. Yet I always respected him as an effective and caring leader who knew how to negotiate better than anyone I've ever met, and even more importantly knew how to get things done. Over time, he came to be an important mentor and dear friend.

Among the subjects that Don and the Colonel had thought carefully about was bushmeat, and it would be a central subject for the work we'd do in central Africa. *Bushmeat* is another word for wild game, although historically the term tends to refer to wild game in tropical locations. In reality, when my friends hunt and eat venison in their yearly New England ritual, they're eating bushmeat. And when I visit my favorite seafood place in San Francisco—Swan Oyster Bar—the living sea urchin they carve open and serve me in the shell is also bushmeat. Yet as we learned in chapter 2, from the perspective of microbes not all bushmeat is created equal.

When we started our work in Cameroon, the overriding objective was to understand why HIV in central Africa was so diverse compared to the fatal but genetically bland and homogenous cosmopolitan versions of the virus that hit most of the world. The idea was to sample HIV from people throughout rural regions and hopefully explain why so many different genetic variants of the virus existed in this part of the world. All of the evidence pointed toward this region as the place where HIV began, but why did it remain so diverse twenty years after the pandemic had exploded?

To answer the question, we teamed up with some of Don's former colleagues at WRAIR (Walter Reed Army Institute of Research), where he had spent most of his career. I remember first meeting the dynamic duo—Jean Carr and Francine McCutchan—at their unremarkable office space in unremarkable Rockville, Maryland. But there was nothing at all bland about the work they'd done.

Over the five years before I met them, the pair had revolutionized the study of HIV by creating methods to sequence entire HIV genomes and systematically study where their various genetic bits and pieces had come from. Prior to their work, people had largely stitched together smaller pieces of genetic information to get a picture of the sequence of the entire virus. Carr and McCutchan had come up with a way to pull the entire ten thousand bits up in one fell swoop. This permitted them to dive into the history of the different genes that made up the viruses.

Since HIV recombines, or has the capacity to mix and match genes among different strains, they needed to form a new set of analytical tools to understand which bits fit together and from where each bit had descended. They were practicing virus genealogy. But instead of piecing together the ancestry of a Scandinavian monarch, they were determining the parental strains of particular HIV viruses and mapping them globally to try and reconstruct the course of the pandemic—plotting out a map of how HIV had spread and mixed.

Along with a dedicated team of local scientists, the Colonel and I would work over the next few years to try to sort out the causes for the intense genetic diversity of HIV in central Africa. Basically, we wanted a snapshot of what HIV looked like before it went global. We started by setting up shop in rural villages throughout Cameroon. The work in the villages was coordinated by Ubald Tamoufe, a soft-spoken and highly meticulous engineer turned biomedical scientist, who still coordinates our joint work in central Africa. We didn't pick just any rural villages. In order to avoid capturing the relatively boring garden-variety

Dr. Don Burke (far left) with (L-R): Dr. Inrombé Jermias, Major Wangmene, Dr. Nathan Wolfe, Ubald Tamoufe, in Cameroon. (*Ubald Tamoufe*)

strains of HIV that now spread throughout the globe and even in regions like Cameroon where the pandemic began, we picked isolated villages found where the roads end.

To say these places were challenging to get to only hints at the complex logistics required to obtain the high-quality specimens Carr and McCutchan needed. These were some of the most remote regions of central Africa. Among their incredible stories comes one from a beloved project driver, Ndongo, who like the Colonel, was largely referred to by his rank, sergent-chef, rather than his proper name. Sergent-Chef once had to abandon his car on one side of a river, then crossed it by small canoe to help our team get specimens from the small village of Adjala in the far southeast of the country.

Sergeant-Chef (L) with GVFI team in Cameroon. L-R: Sergeant Chef Ndongo, Ubald Tamoufe, Alexis Boupda, Ngongang. (Jeremy Alberga)

Obtaining the specimens from these incredibly remote locations held out particular challenges and frustrations but also wonderful experiences. During one particularly memorable visit to a rural village, this one in DRC, I spent the day with hunters in the forest. Upon my return, I learned that a baby boy had just been born to a woman in the village and that they wanted to honor me by giving the child my name. Since one of the villagers had heard me referred to as Docteur Natan (French for "Dr. Nathan," as I'm sometimes called in that part of the world), that was the name they chose for the boy. Not "Nathan" but "Doctor Nathan." The expert research logistician, Jeremy Alberga, who has kept our administrative, logistical, and financial operations organized over the years, joked that the name would decrease the need for burdensome higher education. The boy was already a doctor.

But what exactly were the specimens we were collecting? To start with, we needed blood. From the people who enrolled in our studies, we collected two tubes of blood using high tech tubes that would allow us to easily separate the different parts of blood when we got back to the laboratory in Yaoundé. As for the animals, at least to start with, we worked using a simple but innovative approach, a method developed by Mat LeBreton.

When I originally met Mat in Yaoundé, he was just completing a monumental survey of snakes in Cameroon. Interestingly, he'd done the majority of his sampling by leaving pots of formalin preservative in villages across Cameroon. Since humans throughout the world kill snakes when they find them, he simply asked them to put the dead snakes in the formalin pots, which he'd collect from time to time to study their distribution and diversity. As we talked, we realized that a similar approach could be used to easily collect thousands of specimens from animals. We could adapt the filter-paper techniques I'd learned from Janet and Bal Singh in Malaysia years ago and simply pass out the baseball-card-sized sampling papers to hunters and let

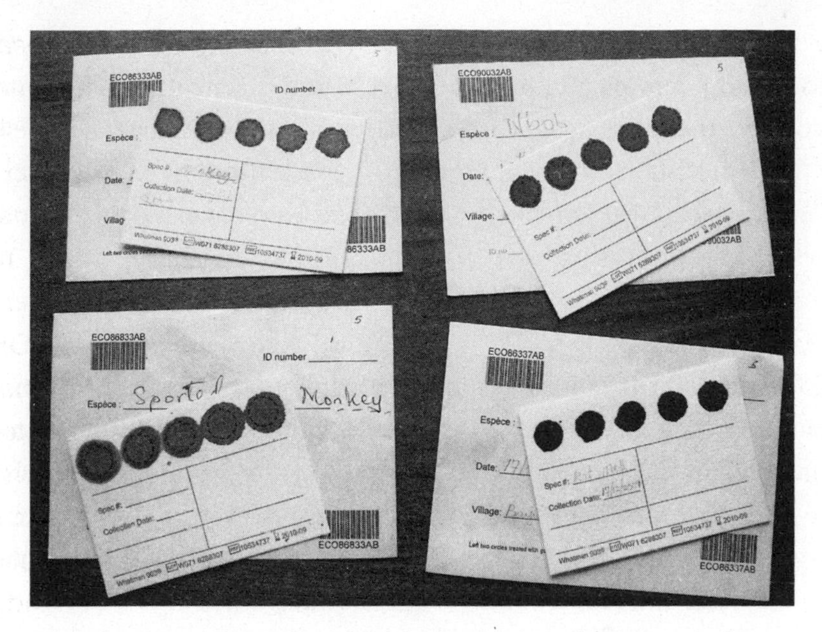

Filter papers used to collect blood samples from bushmeat and other animals. (*Matthew LeBreton*)

them collect specimens whenever they came into contact with blood. The technique proved incredibly successful, and we now have among the most comprehensive wild animal blood collections in the world.

In addition to the challenges of simply getting to tough places to collect specimens, we had the difficulty of communicating our intentions to potential participants in our studies. Gossip and rumors abound in these small villages, and the villager's range of speculations about nefarious purposes for the blood we needed from them was broad. Fortunately, among the excellent staff were some of the most talented communicators I've had the good fortune to work with.

Particularly notable was Paul Delon Menoutou, who had spent years as the chief health correspondent for Cameroonian radio and television, joining us upon his retirement. In many of the rural villages where we worked, people had never had access to television and didn't recognize his face, but when he began to speak, they instantly knew his voice. As a trusted and amazingly talented communicator on health issues, he helped us to ease our way into these communities, which could otherwise resist answering the scientific questions we asked. He also helped convey critical health messages that were a fundamental part of our mission.

Over the first few years in Cameroon, we managed to build a functional laboratory in an amazing century-old building that hailed from the German colonial period in the capital of Yaoundé. We also established connections to seventeen rural villages in fascinating, biologically diverse parts of the country. The high-quality frozen specimens we obtained held clues to the problems of explaining HIV diversity and, as we'd find, much more.

By the time they got into the Rockville laboratories, the specimens had moved thousands of miles by road and air yet remained frozen and viable for testing. I spent some time working in the lab myself, anxious to see exactly what was in the specimens. However, much of the heavy lifting of characterizing the viruses in the specimens was left to McCutchan and Carr and their capable lab teams.

In the end, they found remarkable diversity in these HIV specimens. Twelve of the villages we worked in had completely unique forms of HIV—viruses that were patched together of different sorts of HIV variants, ones that had never been seen together before. In nine of the villages, there were two or more of these unique forms of HIV. Our conclusion was that these locations likely showed what HIV looked like prior to its global spread. Essentially, following the entry of the virus from chimpanzees in the early twentieth century, it likely maintained itself in small villages like the ones we'd studied. Over time, as the virus changed, the newly diverged forms came into contact with each other, shuffling their genetic information and producing an incredible assortment of genetic novelties. Only some of these strains would win the microbial lottery and spread. The rest would remain interesting viruses near the place where their ancestor viruses continued to live in wild chimpanzees, staying put but almost certainly causing disease like their more prolific kin.

While in these rural villages, we did more than just collect specimens to answer our questions about HIV diversity. We also looked into the ways that people interacted with wildlife, a study coordinated at the time by Adria Tassy Prosser, an anthropologist and epidemiologist now based at the Centers for Disease Control in Atlanta. We learned that people in these rural villages had an incredible and intimate level of contact with wild animals. The process of butchering involved direct contact with virtually all of the blood and body fluids that viruses call home. As we expected, people who were engaging in hunting and butchering were at the front line of viral transmission from animals to humans. As we worked in these villages, I became convinced of the potential for these populations to serve as sentinels to allow us to monitor viral chatter. Bushmeat and human contact with it became an almost obsessive focus for me.

But my first real encounter with bushmeat was not in a village. It was at Colonel Mpoudi's house. One of the things the Colonel and I did well together was eat. It was something we'd do in countless villages and cities across the central African region. When you eat dinner at the Colonel's house, you can always expect something special. Along with his gracious wife, Evelyne, herself a successful official of the Ministry of Education, the Colonel has an incredible capacity to make people from any country or class feel comfortable. His dinners in recent times have come to include performances by an excellent local singer—a prominent and vocal member of a group of people living with HIV/AIDS. But whatever the crowd or entertainment, flowing champagne and food are always among the memorable features of the evening. And among the eclectic foods served are always local wild game delicacies. Perhaps my favorite at Chez Mpoudi is porcupine, which tastes a bit like rabbit.

People, wherever they are, traditionally eat wild animals. And while the conservation implications of the consumption of wild animals are important, it's also important that we don't demonize the people that consume these animals to survive. If we could snap our fingers and instantly provide access to high-quality protein sources that didn't involve contact with wild animals, that would be best. It would aid in the conservation of some of the most important endangered species, and it would decrease the prevalence of pandemics. But the problem runs deep.

During the past twelve years, I've worked with many hunters throughout central Africa and in Asia. While illegal commercial hunting exists and must be eliminated, the majority of animals that are hunted in the regions where we work are hunted to provide basic food to needy families. It is subsistence, not entertainment. Hunting is hard work. It requires tremendous energy for a fairly modest outcome in calories.

While many of the hunters we work with are excellent at hunting and some even enjoy it, most would likely choose a cheap and nutritious form of protein that didn't involve hours tracking through incredibly hard-to-negotiate landscapes—fish, for example. I remember my encounter with a man who was headed to his village, carrying a monkey he'd hunted on his back. One of my first thoughts when looking at the bloody and battered animal was how unfortunate it was to lose such a beautiful and important part of our planet's wild heritage. But I also saw that the man was wearing flip-flops and ragged clothing and was sweaty and dirty from a whole day in the forest. He certainly was doing this for his livelihood and not for sport! Subsistence-level hunters are not the enemies and, as we'll explore

Hunter carrying his bushmeat, Cameroon. (*Nathan Wolfe*)

further in chapter 12, the solution is to work with them rather than against them.

As we pushed forward with our work to characterize the diversity of HIV among these rural hunting communities, we also began what would become a main focus of my work over the past ten years—to discover completely novel viruses jumping into these highly exposed peoples. To do so, we approached one of the world's top laboratory teams for discovering novel retroviruses, the broader family of viruses that includes HIV—the CDC's Retrovirology Branch.

The CDC team included Tom Folks and Walid Heneine, two of the world's leaders in retrovirology, but the person I'd spend most of my time working with was Bill Switzer. Bill has a youthful appearance that belies his actual age and a mellow demeanor that masks his relentless drive to chart the evolution of some of the most interesting viruses of our time. Whether face-to-face or by phone, Bill and I would spend the next ten years working together on an almost daily basis to assess what viruses besides HIV had jumped into those hunting populations.

My first major discovery with Bill was of an ominously named virus, the simian foamy virus (SFV). SFV received its name because of the way it kills cells. When you look at a culture infected with the virus, the cells die and bubble up, leading to a foamy appearance under a microscope. It's a virus that infects virtually all nonhuman primates. And since each primate has its own particular version of the virus, it provides a great model for comparison. By sequencing the viruses, if we were then to find one in humans we'd know exactly what animal it had come from.

Interestingly, humans have no indigenous foamy virus. Bill and his colleagues showed some years ago that foamy virus has the unusual feature among viruses of cospeciation. In other words, the common ancestor of all living primates some seventy million years ago had a foamy virus, and as the various branches of the primate tree speciated over time, the virus passed along. The amazing result is that the evolutionary tree of foamy viruses and the evolutionary tree of primates are virtually identical. SFV may very well have been one of the viruses we lost during the pathogen bottleneck discussed in chapter 3.

When Bill and I and our colleagues started the work with primate foamy viruses we already knew that they could theoretically infect humans, as a few lab workers had previously acquired the virus. But we had no idea if this occurred under natural settings. We were surprised and quite excited to find that it did. I remember well the exact day it became clear. We were working together in Bill's lab, and I went downstairs to get the images of a lab test called the Western blot, which shows whether or not individuals have produced antibodies against, in this case, the simian foamy virus. Bill came down that day to help me interpret the images. As soon as we saw the results, it was obvious that some of our study participants had been infected. I remember Bill and I looking at each other with equal parts shock and excitement. In a tangible way the work over the

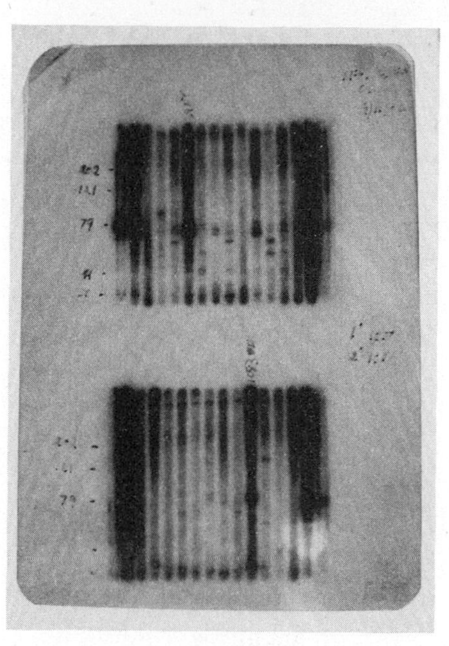

Western blot showing the first evidence of antibodies against simian foamy virus in hunters. *(Nathan Wolfe)*

past years changed dramatically at that moment. To this day I have a framed copy of the Western blot on my wall.

On the one hand, there was relief—the research had succeeded. But there was also a sense of foreboding for us—retroviruses, viruses from the class that had produced HIV, were crossing into humans. And if we were seeing it within the first few hundred hunters we'd studied, it was by no means rare.

Over the coming months we saw that in fact a number of the people in our study who had reported hunting and butchering nonhuman primates had been exposed to SFV. More amazingly, some of the exposures had gone on to become long-lasting infections. After finding evidence that these individuals produced antibodies to the virus, we tried to obtain actual SFV genetic sequence, and what we saw surprised us. We found multiple people who were infected with strains of SFV from primates, ranging from the DeBrazza's guenon, a small leaf-eating monkey, to the massive lowland gorilla. To our great satisfaction, we found that the results of our behavioral surveys matched. The gorilla SFV, for example, came from a man who had reported hunting and butchering gorilla meat. While many of the people in our survey had exposure to primates, few participated in the dangerous and highly specialized hunting of gorillas. The link was a smoking gun—the gorilla hunter had acquired the virus while hunting or butchering his prey.

The finding provided both a sense of adulation and fear. Most virologists would be lying if they said they didn't enjoy finding something completely new. It had taken us years of hard work to line up the funding, find the local collaborating scientists who knew how to accomplish the work, set up a lab in central Africa, establish village outreach, collect specimens, store and ship them through the complexities of international agreements, and conduct the complicated laboratory work necessary to find an actual virus. The results showed that our system worked and that we were right in guessing that high levels of exposure to animals would lead to infection with novel viruses. Yet, the first evidence that new retroviruses were moving into humans also suggested that people's faith in the existing public health structures—that they would inform us when novel viruses were moving into humans—was misguided. We were only beginning to see just how misguided it was.

In the following year, we went on to study yet another group of retroviruses, the T-lymphotropic viruses (TLVs). SFV was a virus with no real human precedents. Before our work, only a handful of laboratory workers had been infected, so determining how much the virus was likely to spread and cause disease—and its potential to become a pandemic—was unclear. Not so for the TLVs. It's long been known that humans around the world are infected with two different varieties of TLV—HTLV-1 and HTLV-2; in fact, some twenty million people have these viruses. While some individuals can be infected without disease, many get sick from illnesses ranging from leukemia to paralysis. These viruses have pandemic potential. Clearly, if completely novel TLVs were moving from animals to humans, public health officials should know about it. Our results from SFV suggested this was a real possibility.

Going into the study, Bill and I knew that each of the two varieties of human TLV came from primates—just as HIV had. We also knew that another group of TLVs existed in primates that hadn't yet been found in humans—the Simian T-lymphotropic Virus 3, known as STLV-3—so we began there. We screened the samples carefully, and as predicted, we found it—a virus infecting human hunters that was clearly unlike HTLV-1 and HTLV-2 and fell squarely with the viruses in the STLV-3 group. This was an important scientific finding for us. STLV-3 had the potential to cross into humans and was on the move. Even more surprising was an entirely new human TLV found in a single individual from eastern Cameroon—a virus we called HTLV-4.

The combination of finding a number of new SFVs in people exposed to primate bushmeat in central Africa and two entirely novel TLVs in the same population changed the way that we thought about our work. While it was theoretically clear that people exposed to a wide range of wildlife would acquire microbes from these animals, we didn't know at the start whether monitoring those populations was practical or what such a system would look like. As we began the long and plodding work to determine the extent to which the new SFVs and TLVs were spreading and causing disease (work that continues to this day), our thinking opened up. We began to seriously consider that monitoring people highly exposed to wildlife could be a globally deployed system to capture viral chatter.

In 2005 I took a long shot. I applied to an unusual program sponsored by the National Institutes of Health (NIH), the largest government funder of biomedical research in the world. The NIH had supported my work in the past, but the world-class institution didn't perfectly match the work I hoped to do in the future. While the NIH has a broad ranging program, it does not distribute its resources evenly. The NIH specializes in funding laboratory research rather than field research. It focuses its energy largely on research in more reductionist cell biology—work that focuses on very clear hypotheses that provide very clear yes or no answers. A program to spearhead a brand-new global monitoring effort to chart viral chatter and control pandemics was not something that would normally be supported. Yet in 2004 the NIH began a completely new program—the NIH Director's Pioneer Award Program—aimed at sponsoring innovative research not normally supported by the NIH mission. The program gave grantees $2.5 million and five years to do largely whatever they felt was necessary to advance their scientific objectives. In the fall of 2005 I was among the fortunate individuals to get the award.

At this point, the pieces were beginning to fall into place. Certainly $2.5 million was nowhere near what would be needed to roll out a global monitoring system, but it was a good start. It allowed me to begin truly thinking about which key viral hotspots around the world needed the most urgent monitoring. Some key regions came to mind right away. The work with Jared Diamond and Claire Panosian had shown that Africa and Asia provided the lion's share of our major infectious diseases. Those would be the places to start.

In the coming years, along with my team and a stunning range of local collaborators, I would take the model we'd developed in Cameroon and begin to deploy it in a number of other countries in central Africa. With the help of dedicated field scientists like Corina Monagin, who has become expert at making field sites in sensitive and difficult areas function, I'd renew collaborations from my years in Malaysia and begin to work with new sets of colleagues to establish programs in China and Southeast Asia. We'd set up the beginnings of a system to capture global viral chatter. Along with a growing number of colleagues worldwide, we would push ourselves to ask how we could best find new viruses. How could we capture a much higher percentage of the new viruses killing humans and infecting animals?

In the coming chapters, we'll explore the results of this work. I'll also discuss some of the cutting-edge tools employed to improve our ability to detect pandemics before they spread. While the threats associated with pandemics are large and growing, so too are the approaches and technologies to address them.

➤ 10 ➤

MICROBE FORECASTING

It was a large city. And it was hit hard. The first cases emerged in late August, and the victims suffered terribly. The earliest symptoms were profuse diarrhea and vomiting. They experienced severe dehydration, increased heart rate, muscle cramps, restlessness, severe thirst, and the loss of skin elasticity. Some of the cases progressed to kidney failure, while others led to coma or shock. Many of those who came down with the disease died.

Then on the night of August 31, the outbreak truly broke. Over the next three days, 127 people in a single neighborhood died. And by September 10 the number of fatalities would reach 500. The epidemic seemed to spare no one. Children and adults alike were killed. Few families did not have at least one member who came down with the disease.

The epidemic led to intense panic. Within a week, three-quarters of the neighborhood's residents fled. Stores closed. Homes were locked. And you could walk down a formerly bustling urban street without seeing a single person.

Early in the outbreak, a forty-year-old epidemiologist began an investigation to determine its source. He consulted community leaders and methodically interviewed families of the victims and made careful maps of every single case. Following his hunch about a waterborne disease, he studied the sources of the community's water and determined that it came from only one of two urban water utilities. He conducted microscopic and chemical analyses of specimens from the water system, which proved inconclusive.

In his report to the responsible officials, he presented his analysis and concluded that contaminated water was to blame. Despite the lack of definitive results from the analyses, the mapping of cases strongly supported his conclusion that one particular water outlet was the source of the outbreak. He recommended shutting down the water supply, and the officials agreed. And while the outbreak may have already been in decline because of the mass exodus, that investigation and water closure proved pivotal.

What was unusual about this outbreak was not the procedural investigation that followed. Modern epidemiologists in countries throughout the world conduct exactly this kind of investigation regularly. They enlist the help of local leaders, study the distribution of cases, conduct analyses on potential sources, and then often argue with officials as to the best course of action. What was unusual was that the outbreak was in 1854—before the field of epidemiology existed.

As you may have guessed, the investigator responsible for cracking the outbreak was none other than John Snow, the now famous London physician and clergyman considered one of the founders of contemporary epidemiology. The culprit was, of course, the bacteria *Vibrio cholerae*, or cholera. By finding that water was the source rather than "foul air," Snow contributed to the modern germ theory of infectious diseases—that communicable diseases are caused by microbes. To this day, you can see a replica of the famous Broad Street pump that Snow identified as the source of the 1854 outbreak, in Soho, London.

It seems intuitive to us today, but the way that Snow used interviews, case identification, and mapping to chart the origin of the Broad Street cholera outbreak of 1854 was revolutionary in its time. While maps had certainly been used extensively prior to 1854, the map he made of Soho is considered the first of its kind, not only in epidemiology but also in cartography. He was the first to utilize maps to analyze geographically related events to make a conclusion about causality—namely, that the Broad Street pump was the source of the outbreak. By doing so he has been credited with using the first geographic information system, or GIS, a now commonly used cartographic system for capturing and analyzing geographic information.

In contemporary GIS, layers of information are added to maps like Snow's to provide depth of geographic information and to suggest patterns of causality. While Snow's map included streets, homes, locations of illness and water sources, a contemporary version could include many more layers—genetic information from cholera specimens collected in different locations, dimensions of time that track changes spatially with an added weather layer or social connections between the individuals in the various homes.

Modern GIS is among a range of contemporary tools that is radically changing the way that we investigate outbreaks and understand the transmission of diseases. When used in a coordinated and comprehensive way, these tools have the potential to fundamentally change the way that we monitor for outbreaks and stop them in their tracks.

We now have multiple scientific and technical advantages that Snow lacked in the mid-nineteenth century. Among the most profound is that we have significantly improved our capacity to

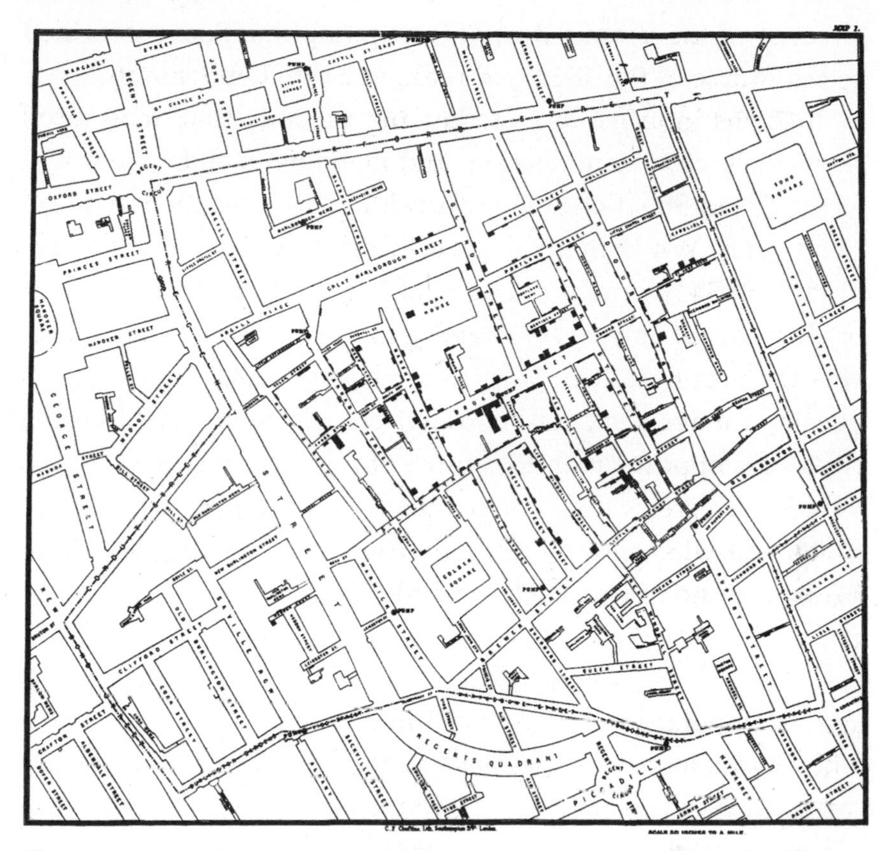

The map of London used by John Snow to find the source of the cholera outbreak.

catch the bugs we're chasing and to document their diversity. The revolution in molecular biology, in particular the techniques for capturing and sequencing genetic information, has profoundly changed our ability to identify the microbes that surround us.

Miraculous but now standard techniques like the *polymerase chain reaction* (PCR), which resulted in the Nobel Prize for its discoverer Kary Mullis, allow us to snip out tiny pieces of genetic information from microbes and create billions of identical copies, whose sequences can then be read and sorted out according to the family of microbes to which they belong. Yet standard PCR requires that you know what you're looking for.

Dr. John Snow, 1856. (*Wellcome Library, London*)

If, for example, we want to find an unknown malaria parasite, we can use PCR designed to identify malaria-specific sequence, since all malaria parasites have genetic regions that look similar enough to each other. But what if we don't know what we're looking for?

In the early 2000s, intent on finding unknown microbes, a bright young molecular biologist, Joe DeRisi, and his colleagues adapted an interesting technique developed by DeRisi's doctoral adviser, Pat Brown, a Stanford biochemist. The *DNA microarray chip* consisted of thousands of tiny bits of distinct artificial genetic sequence distributed in an orderly fashion across a small glass slide. Since genetic information sticks to its mirror image sequence, if you flush solution from a specimen containing genetic information across a slide like this, the bits that match the designed sequences on the slide will fuse. You can then determine what was in the specimen by determining which of the sequences on the slide trapped their natural siblings. The technique had already provided thousands of scientists with a new way of characterizing the bits of genetic information that flow through living systems by the time DeRisi got his hands on it.

Prior to DeRisi's innovation, the microarray chips had been used primarily to help determine the internal workings of the genes of humans and animals, but DeRisi and his colleagues realized that the technique could be modified to create a powerful viral detection system. Instead of designing the chips with bits of artificial human genetic information, he and his colleagues designed chips with bits of *viral* genetic information. By carefully reviewing the scientific databases for genetic information on all of the viruses known to science, they crafted chips that had bits of genetic information from a whole range of viral families lined up in neat rows. If they introduced genetic information from a sick patient, and it contained a virus with a sequence similar to one on the chip, the sequence would be trapped and—bingo!—we'd know the bug we were dealing with.

The *viral microarray*, as these specialized chips became known, have proliferated and spread to labs throughout the world. They've helped quickly identify the microbial villain responsible for new pandemics, like the coronavirus that causes SARS. Yet they are not perfect. These chips can only be made to capture viruses from families of viruses already known to science. If there are groups of viruses out there whose sequences we are completely unaware of, and there certainly are, then we have nothing with which to engineer the chips. Truly unknown viruses would slide right by.

Within the past few years, viral microarrays have been supplemented with a series of bold new genetic sequencing approaches. New machines churn out mammoth amounts of sequence data from specimens—amounts of sequence that previously would have been prohibitively expensive or time consuming. These machines are permitting an entirely new form of viral discovery.

Rather than look for particular bits of information, the approach is to take a specimen—say a drop of blood—and sequence every bit of genetic information it contains. Technically, it's more complicated than that, but the result is similar to what you'd expect. We are approaching a moment when we will be able to read every single sequence within a given biological specimen. Every bit of DNA or RNA from the host specimen, and critically, every bit from the microbes that are riding along with them.

One of the central problems becomes the bioinformatics—how to sort through all of the billions of bits of information that are produced by these incredible technologies. Fortunately, in an enlightened move, scientists at the NIH picked up and nurtured an electronic repository of sequencing information developed at the famed Los Alamos National Laboratory and now called GenBank. Since scientists are required by funding sources and journals to submit sequences to GenBank prior to submitting academic papers, we collectively contribute billions of bits of genetic information each year. GenBank right now holds over a hundred billion bits of sequence information. And it's growing rapidly. When a new sequence is identified from a sequencing run, it can be rapidly compared electronically to what's in GenBank to see if there's a match.

In late 2006 and early 2007 these techniques were used to good effect. In early December 2006 the organs of a patient who had died of a brain hemorrhage in Dandenong hospital in Australia were harvested for transplantation. A sixty-three-year-old grandmother received one of the kidneys, another unnamed recipient received the other kidney, and a sixty-four-year-old lecturer in a local university received the man's liver. By early January all three had died.

The local hospital and collaborating labs looked for all of the usual suspects. They utilized PCR and tried to grow up the microbe on culture media. They even tried one of the viral microarrays, to no avail. A virus was only found when the specimen was subjected to massive sequencing. The team that found it, led by Ian Lipkin, a world-class laboratory virologist at Columbia University, had to sort through over a hundred thousand sequences to find the fourteen sequences belonging to the mystery virus. Truly a needle in a haystack! The mystery virus ended up being in a group of viruses called arenaviruses that often live in rodents. Without massive sequencing, the virus would not likely have been found.

But while identifying what's actually in a small new outbreak is vital, it's only the beginning. As we get better and better at understanding what's out there, we will have to start asking a tougher question: where is it going? Will it become a pandemic?

There are three primary objectives to the emerging science of pandemic prevention:

1. We need to identify epidemics early.
2. We need to assess the probability that they will grow into pandemics.
3. We need to stop the deadly ones before they grow into pandemics.

The viral microarray and sequencing techniques give us a snapshot of what is causing an epidemic, but more is needed to assess the possibility that a new agent in a limited outbreak has the right stuff to go pandemic. This is exactly the objective of a new program being developed by DARPA, the U.S. Department of Defense's Advanced Research Projects Agency. DARPA has had a stunning impact on the contemporary world of technology, including sponsoring early research that has contributed in substantive ways to the development of modern computing, virtual reality, and the Internet itself.

DARPA is developing a program called Prophecy, whose

objective is to "successfully predict the natural evolution of any virus." Prophecy seeks nothing less than to use technology to predict where an outbreak will go by combining it with the support of a team of local on-the-ground experts in hotspots around the world. Predicting the future trajectory of a virus seems like science fiction, but DARPA does not shy away from high-risk/high-payoff ideas, and Prophecy falls clearly in that mold. Fortunately, what we know about pandemics and the technologies available today bring the objectives it seeks within the realm of possibility.

Cutting-edge experimental virologists like Raul Andino, at the University of California, San Francisco, works to determine rational predictions of the evolution of viruses. Viruses reproduce rapidly, so any viral infection, even if it's the result of infection with a single viral particle, will rapidly develop into a swarm,* a group of viruses, some identical, but mostly mutants differing in one way or another from the parental strain that created them. By documenting and studying the way that the overall viral swarms respond to different environments, Andino and his colleagues have worked to develop rational strategies for the production of vaccines that use live viruses, a subject we will return to in chapter 11. He also hopes to use the same information to determine the boundaries within which a swarm can evolve. Swarms can't go in every direction, and getting a sense of what a swarm is composed of will help us get a sense of what it can evolve into.

Another scientist working to change the ways we can forecast microbial evolution is not a microbiologist at all but rather a physics-trained bioengineer. Steve Quake, an awardee of the same NIH Pioneer Program that has funded my own research, develops technology that permits us to study and manipulate

* Such swarms are also referred to as viral quasispecies in the scientific literature.

life in surprising and incredibly useful ways. In the past ten years this jeans-wearing ski bum has spun off multiple companies, developed handfuls of patents, and published scores of papers in some of the highest-ranking journals—all while maintaining a successful teaching program at Stanford University. Among the useful innovations coming from Quake's group are *microfluidic platforms*. Essentially, he's produced entire laboratories on small laboratory chips.

In one particularly notable application, he's taken the tedious and complex work of cell culture, where cells from mammals and other organisms are grown under laboratory conditions, from the bench to the chip. The chips he and his team have created, just a few centimeters long, house ninety-six separate compartments where cells grow for weeks at a time and can be carefully measured and manipulated. While there are many applications for having cell culture on an automated and compact chip, one of them is the speed and efficiency for evaluating new viruses from large numbers of specimens. It's not difficult to imagine a chip-based system that quickly tells us in what kind of cells a new agent can survive and therefore how it's most likely to spread (e.g., by sex, blood, sneezes, and so on).

When we see an outbreak, there are a number of questions we'd like to have answered. First, what's the microbe behind it? Techniques like viral microarrays and high throughput sequencing are increasing the speed at which we can identify new agents and also helping us to find things that we'd have missed through older techniques. But once we've identified a microbe, we want to know where it's going. We'll return in chapter 12 to a vision of what the ultimate pandemic prevention system will look like, but it would certainly involve approaches like those developed by the Andino lab to assess the potential evolutionary directions that a virus can take. And the tools that Quake's group has developed might one day form a set of high-speed chips that quickly evaluate how it's likely to spread.

• • •

Modern information and communication technology provides us with another set of tools that does something distinct and complementary to the biotech advances discussed above. In fact, some of this technology is sitting in your pocket as you read this sentence.

In one of our research sites in southwest Cameroon sits a rubber plantation called Hevecam, where we conducted an experiment. This experiment represents one of the exciting new trends in public health. And it's all based on simple cell phones.

In Hevecam, a plantation with nearly a hundred thousand inhabitants, when individuals get sick they go to a local clinic. If they're sufficiently ill, they then move from that local clinic to the referral hospital in the center of the plantation. Yet traditionally there has been no good way for the referral hospital to monitor what's happening in the local clinics. A few years ago Lucky Gunasekara, who now heads up our program on digital epidemiology, and his partners at the nonprofit FrontlineSMS:Medic that he co-founded, set up a simple system based on text messages to allow the referral hospital to monitor what was occurring in the local clinics. By simply texting a series of preset codes, the vast majority of vital clinical information could be communicated up the medical hierarchy clearly, instantly, and efficiently. Using predetermined codes and simple text message forms, the local clinics could rapidly inform everyone else of how many cases of malaria, diarrhea, and other illnesses they were seeing.

Simple technologies can have dramatic impact. With a few simple techniques, medical conditions at Hevecam could be monitored not only in the referral hospital but also remotely over a web dashboard for anyone with appropriate access. By allowing local clinicians or patients themselves the capacity to communicate, information can be accumulated, organized, and analyzed, leading to a much more rapid and localized sense of what's going on during a health emergency.

Something just like this occurred during the earthquake in Haiti in 2010. Immediately after the earthquake, organizations like Ushahidi* set up short, free codes to which people could text "help" messages. They then turned to the local DJs who, along with popular word of mouth, publicized the numbers. Amazingly, when the dust cleared, the statistical analysis of the text message distributions mapped accurately onto high-resolution aerial imagery of damage. Effectively, people's text messages gave highly informative clues as to where the greatest damage occurred. More importantly for those in Haiti, the messages saved lives, with the critical information transmitted to the heroic rescue workers on the scene.

Similar systems have been used during outbreaks, such as the cholera outbreak in Haiti in the fall of 2010. The ultimate hope is that outbreak detection can be crowdsourced, with small bits of information provided by sufferers that converges into a real-time picture of the beginnings of outbreaks and their subsequent spread. The short codes are only the start. As more and more countries adopt electronic medical records, people around the world will increasingly link to them directly by reporting their health complaints from their phones. This information will not only provide more efficient medical care to individuals who report illness—when analyzed across large numbers of users, it will allow more rapid and sensitive detection of health anomalies. Eventually, response systems can be developed to recognize unusual clusters of health complaints that

* Ushahidi is a pioneering nonprofit technology company that works to improve collection, visualization, and mapping of information. The word *ushahidi* means "testimony" in Swahili, and the company was started after the postelection violence in Kenya in 2008 to help consolidate and map reports of violence.

signal the beginnings of an epidemic. With that, the age of digital epidemiology will truly have begun.

One of the critiques of using text messaging as an early indicator of disease spread is that even under the direst circumstances not everyone will text. But there are ways to use cell phones that don't require their users to do a thing.

At the moment I'm writing this sentence, over 60 percent of the world's population has been planted with automated locating beacons. These beacons provide constantly updating information on exactly where they are. Within the next five to ten years, virtually everyone on the planet will have one. This is not a government plot. It is the mobile phone in your pocket.

Cell phones constantly communicate with cell towers, providing telecom operators with an incredibly rich amount of data about where their customers are, how the customers are connected to each other, and with a bit of interpretation, the social behaviors of their users. These so-called call data records have provided huge data opportunities for the telecoms to understand their clients and sell them more services. But the massive data sets have much more value than sales. This constant flow of seemingly innocuous information could save your life.

The data collected by cellular telephone companies makes us all potential sensors for rapid detection of important human events. This was shown elegantly by Nathan Eagle, a member of the innovative MIT Media Lab and one of the pioneers in applying call data records to generalized problems. Along with his colleagues, Eagle sought to investigate what could be known about an earthquake by mining call data records.

Eagle and his team studied data on calling patterns in Rwanda for three years, including the critical week of February 3, 2008, when a 5.9 magnitude earthquake occurred in the

Lake Kivu region. By establishing a baseline for the frequency of calls, Eagle and his team were able to see telltale clues of unusual calling patterns during the period immediately following the earthquake. They were able to detect the time of the quake through a peak in call numbers. They were also able to establish the epicenter of the quake by using location data from cell towers, placing the epicenter central to the locations of the heaviest call volumes.

The idea that using data derived from cell phones can detect an earthquake in space and time is amazing. It also suggests a range of different applications. Individuals who are ill may have fundamentally different call patterns than those that are not, and call patterns may also alter as a new outbreak spreads. Analyses of call data records alone might not provide perfect early detection of a new outbreak, but combined with other sources of outbreak data from organizations like ours and other health institutions, it might help us chart early epidemic spread.

Cell phones are growing more ubiquitous by the day and will likely be critical tools in helping to detect and respond quickly to outbreaks before they become pandemics. Yet they are not the only technology-heavy solutions being used in the growing field of digital surveillance. In 2009 my colleagues at Google* published a fascinating paper showing that individuals' online

* The Google team that discovered that search trends correlate with actual influenza incidence included Larry Brilliant and Mark Smolinski, formerly of the Google.org Predict and Prevent project. It also included young Google engineers, who through Google policies can devote a percentage of their time to philanthropic or other endeavors. Both Larry and Mark have now joined Jeff Skoll, the entrepreneur, filmmaker, and philanthropist in his new endeavor, the Skoll Global Threats Fund, which focuses on ways to mitigate the threat from some of the most important risks of our time—they include, of course, pandemics.

search patterns also provide a sense of what people are becoming infected with.

With the vast stores of search data kept by Google and US influenza surveillance data collected by the CDC, the team was able to calibrate their system to determine the key search words that sick people or their caregivers used to indicate the presence of illness. The team used searches on words related to influenza and its symptoms and remedies to establish a system that accurately tracked the influenza statistics generated by the CDC. In fact, they did better. Since Google search data is available immediately, and CDC influenza surveillance data lags because of time needed for reporting and posting, Google was able to beat the CDC in providing accurate influenza trends before the traditional surveillance system.

Early data on seasonal influenza, as provided by the Google Flu Trends system, is interesting and potentially important. This early data provides health organizations time to order medications and prepare for different triage needs. But early detection of seasonal influenza is not the Holy Grail. That honor would go to a system that could detect a newly emerging pandemic. Google is now working to extend its influenza findings to other kinds of diseases. As more and more people use search engines like Google, and more and more data is acquired, the hope is that better and better trend analyses will be developed for agents other than influenza. Perhaps at some point a community experiencing the beginning of a pandemic will signal its arrival just by Googling.

The explosion of online social media provides another set of *big data* in which weak but potentially valuable signals of a coming plague may be found. Computer scientists, like Vasileios Lampos and Nello Cristianini from the University of Bristol, have taken a similar approach as the scientists at Google, sorting through hundreds of millions of Twitter messages. Like their colleagues at Google, Lampos and Cristianini used key words to watch trends in Twitter and find associations with influenza statistics, in this case provided by the UK's Health Protection Agency.

In 2009 they tracked the frequency of tweets related to influenza during the H1N1 pandemic and found they were able to track the official health data with 97 percent accuracy. As with the findings by the Google Flu Trends team, this work provides a rapid and potentially inexpensive way to supplement traditional epidemiological data gathering. It also has the potential to be extended to more than just influenza.

While online social media can be scanned to see what people are communicating about, online social networking may provide a richer and subtler range of possible uses. In fascinating recent work, two leading social scientists, Nicholas Christakis and James Fowler, have studied how social networks can inform surveillance for infectious diseases.

In a clever experiment, these two scientists followed Harvard students who were divided into two groups. The first group was randomly selected from the Harvard student population. The second group was chosen from individuals that the first group named as friends. Because individuals near the center of a social network are likely to be infected sooner than those on the periphery, Christakis and Fowler hypothesized that during an outbreak the friend group would become infected sooner than the random, and therefore on average less socially central, group. The results were dramatic. During an influenza outbreak in 2009, the friend group became infected on average fourteen days ahead of the randomly chosen group.

The hope is that social science can identify novel kinds of sentinels to monitor for new outbreaks and catch them early.*

* Social networks are not the only social science approaches to early detection. Another approach is to use *prediction markets*. In the 2004–5 influenza

Determining friends would be time consuming, however—something we could accomplish on a single college campus but perhaps not nationally. Now self-identified friends in massive online social networks may make this task much easier. Online social networks like Facebook, while not designed to help monitor for outbreaks, have created relatively easy-to-monitor systems that can be mined to determine the frequency of illness, identify social sentinels, and perhaps eventually provide predictions for spread of a new agent within a community.

When John Snow created the first Geographic Information System in 1854, he took actions that would seem very logical and straightforward to us today. He took a map, he plotted where sick people were, and he plotted possible sources of contagion. Snow could not have predicted the directions in which his first tentative step would lead or the data that would eventually become available for today's GIS.

In the end it may be that no single data source reigns supreme. If Snow were alive today and investigating an outbreak, he'd want it all. He'd want to know where the sick people were, and he'd be glad to get the data more quickly and easily through text messages or Internet searches. He'd like to know exactly what cases were infected with, down to the very specific microbial genetic strain. He'd seek to use call data records to monitor people's movements in order to track the movement of the disease or where it was seeded. He'd like to know how people were connected socially, and he'd certainly follow individuals who were likely to become infected first or show signs earlier than the rest.

---

season, researchers at the University of Iowa set up a futures market where nurses, pharmacists, and other health workers could trade and make money (in the form of an educational grant) on their sense of what was going on with influenza. The researchers showed that looking at market activity of local experts incentivized to choose correctly can also provide early warning.

You can imagine the ultimate outbreak GIS, or in terms more familiar to Silicon Valley, what Lucky Gunasekara, the head of my data team, calls the ultimate outbreak *mash-up*: a map with layer after layer of critical information—where people are, what they're concerned about, what they're infected with, where they're moving, and who they're connected to. Developing and maintaining this combined digital and biological mash-up is the precise objective of Lucky's team and something to which we'll return in the final chapter of this book. Ideally, over time the data can be analyzed jointly, the various factors can be trained on actual outbreaks, and all the technology can be weighted optimally to maximize predictive power.

When people ask me whether or not I'm optimistic about the future of predicting pandemics, the answer is always a resounding yes. Given the first two-thirds of this book, you may wonder if my optimism is warranted. A steady wave of interconnectedness among humans and animals has created a perfect storm for new pandemics. That is true. Yet the interconnectedness among humans that now exists through communication and information technology gives us unprecedented capacity to catch outbreaks early, which, when combined with amazing advances in our ability to study the diversity of the tiny life forms that cause epidemics, certainly makes optimism warranted.

What will win out in the end? Will pandemics sweep through the human population destroying millions of lives? Will technology and science ride in to the rescue?

➤ 11 ➤

THE GENTLE VIRUS

All living organisms focus huge amounts of energy on having successful offspring. In humans, this means breastfeeding and constant care of babies for the first few years of their lives. In other organisms, like sea tortoises, the energy is spent not on care for existing offspring but in creating the conditions necessary to successfully launch hundreds of instantly self-sufficient infants—accumulating nutrients to place in eggs, traveling to the right place to lay eggs, and burying eggs in sand to protect them from predators. Whatever they may look like, parents want their kids to succeed, and they deploy a range of techniques to aid them in that objective.

Among the concerned parents out there are wasps. Two families of wasps go to an extraordinary measure to protect their offspring. These wasps, of the braconid and ichneumonid families, lay their eggs on the backs of caterpillar larvae. The eggs then eat the flesh of the caterpillar as they grow. This is actually a fairly common setup on our planet, with thousands of such relationships in existence. There is an evolutionary tension between the caterpillar and the wasp. The caterpillar's

Braconid wasp eggs on a caterpillar larva. (*James H. Robinson / Photo Researchers, Inc.*)

defenses change over time to thwart the wasp eggs, and the wasp eggs develop the capacity to counteract or skirt the caterpillar's defenses, and so on.

In their battle to win this evolutionary arms race, the female braconids and ichneumonids do something not known among other wasps that live in this way: they coat their eggs in a special substance before they lay them on the back of a caterpillar. Slowly, this potent substance kills the caterpillar, leaving the eggs to grow unrestricted on the bounty that remains.

The wasp mothers' truly amazing substance is not a plant toxin or a venom. It's a concentrated dose of virus. This virus, a member of the polydnavirus family, harmlessly infects the wasp but unleashes a range of consequences in the caterpillar. It replicates in the wasp's ovaries and is injected, together with the wasp's eggs, into the caterpillar. The virus returns the favor by suppressing the host caterpillar's immune system and causing severe disease and even death to the caterpillar, thereby protecting the eggs. The wasp helps the virus, and the virus helps the wasp.

Viruses operate along a continuum with their hosts: some harm their hosts, some benefit their hosts, and some—perhaps most—live in relative neutrality, neither substantively harming nor benefiting the organisms they must at least temporarily inhabit for their own survival.

In this chapter we'll shift gears. Rather than discuss the harm viruses can cause, we'll focus on how they can assist us in the battle against infectious and other diseases. The goal of public health should not be to eradicate all viral agents; the goal should be to control the deadly ones.

Perhaps the most profound way that viruses have assisted us in the fight against pandemics has been in the case of vaccines. And there is no better example of this partnership than our relationship with the cowpox virus.

In the late eighteenth century, the noted English scientist Edward Jenner became fascinated with the observation that milkmaids somehow seemed to avoid becoming infected with smallpox. On May 14, 1796, taking a bit of a leap, Jenner inoculated James Phipps, the eight-year-old son of his gardener, with cowpox that he'd scraped from the hand of a young milkmaid named Sarah Nelmes. She had acquired the virus from a cow named Blossom, whose hide you can apparently still see if you visit St. George's medical school in London.

Young James Phipps got mildly sick, a bit of fever and some discomfort but that was all. After James recovered, Jenner went on to inoculate the boy with a small amount of the actual smallpox virus.* The smallpox did nothing. The effect, which Jenner then replicated in others, would go on to be one of the most profound findings in human history. He had developed a vaccine to prevent smallpox, one of the worst scourges of humankind. The discovery is credited by some as saving more lives than any other discovery in history.

The vaccines that were created as a result of Jenner's work eventually led to the eradication of smallpox from the planet. I remember seeing one of the original documents certifying that smallpox had been eliminated. It was in the Johns Hopkins office of D. A. Henderson, who had led the WHO's global smallpox eradication campaign. D. A. had kindly lent me one of his largely unused offices at Hopkins as a staging ground to accumulate the supplies I'd need to start our work monitoring outbreaks in central Africa. I remember thinking to myself about how important eradication was and how it had been accomplished.

We credit the eradication of smallpox to a vaccine. But it's worth examining this further. The vaccine that allowed us this triumph was actually an unadulterated virus that we harnessed and used for our benefit. In fact, even the word *vaccine* itself derives from the Latin term for cowpox, or *variolae vaccinae,*

---

* You may question the ethical decision of Jenner to experiment on a child with an unproven vaccine and then to expose the child to a known deadly disease. Yet while he has been critiqued for it, a more careful examination reveals something quite different. Because the rate of smallpox was relatively high, many adults would likely already have been exposed, making them inappropriate for the study, necessitating a study in children. Also, when he injected Phipps with smallpox it was part of an even earlier form of smallpox vaccination called variolation, in which patients were exposed to small amounts of actual smallpox virus (i.e., variola) in a controlled way to illicit an immune response to protect from natural infection. Variolation killed 1–3 percent, a crazy level by today's standards but much lower than the 30 percent mortality among those who were naturally infected. Considering these factors, and that he also included his own son in these experiments, I think we can probably let Jenner off the hook.

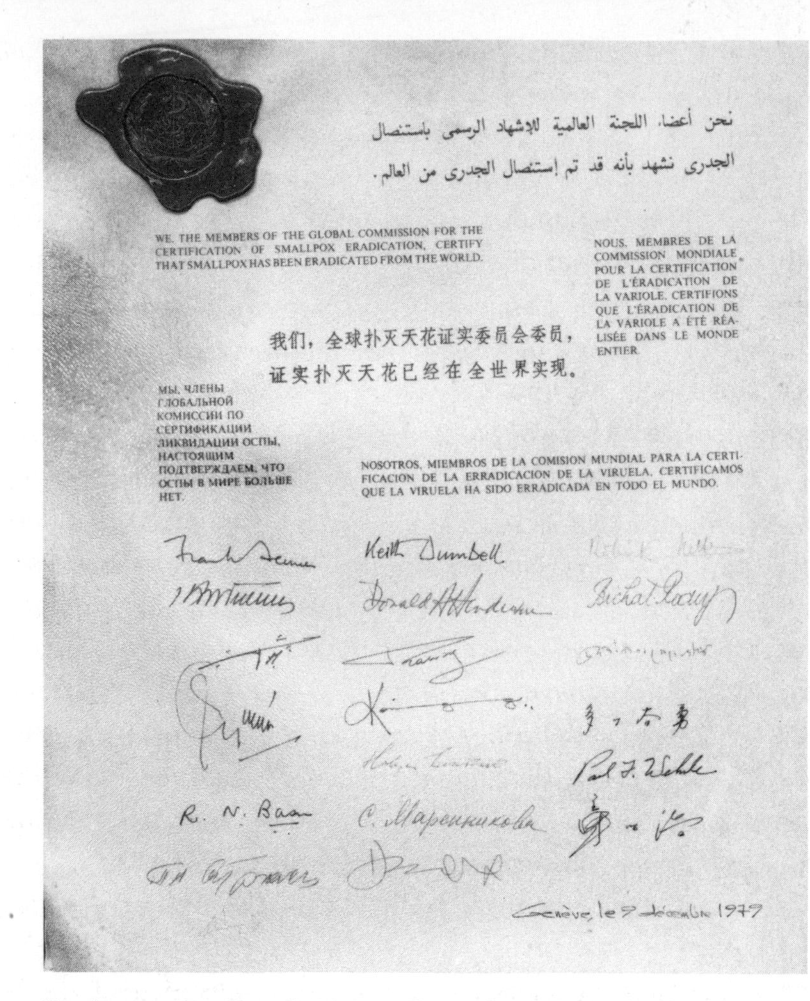

نحن أعضاء اللجنة العالمية للإشهاد الرسمى باستئصال
الجدرى نشهد بأنه قد تم إستئصال الجدرى من العالم.

WE, THE MEMBERS OF THE GLOBAL COMMISSION FOR THE CERTIFICATION OF SMALLPOX ERADICATION, CERTIFY THAT SMALLPOX HAS BEEN ERADICATED FROM THE WORLD.

NOUS, MEMBRES DE LA COMMISSION MONDIALE POUR LA CERTIFICATION DE L'ÉRADICATION DE LA VARIOLE, CERTIFIONS QUE L'ÉRADICATION DE LA VARIOLE A ÉTÉ RÉALISÉE DANS LE MONDE ENTIER.

我们，全球扑灭天花证实委员会委员，
证实扑灭天花已经在全世界实现。

МЫ, ЧЛЕНЫ ГЛОБАЛЬНОЙ КОМИССИИ ПО СЕРТИФИКАЦИИ ЛИКВИДАЦИИ ОСПЫ, НАСТОЯЩИМ ПОДТВЕРЖДАЕМ, ЧТО ОСПЫ В МИРЕ БОЛЬШЕ НЕТ.

NOSOTROS, MIEMBROS DE LA COMISION MUNDIAL PARA LA CERTIFICACION DE LA ERRADICACION DE LA VIRUELA, CERTIFICAMOS QUE LA VIRUELA HA SIDO ERRADICADA EN TODO EL MUNDO.

Genève, le 9 décembre 1979

Parchment signed at Geneva on December 9, 1979, by the members of the Global Commission for Certification of Smallpox Eradication. (*World Health Organization*)

where *variolae* means "pox" and *vaccinae* means "of cows." In other words, at its very heart, the concept of a vaccine is the productive use of one virus to fight another.

Because cowpox is close enough to smallpox that it leads to immunity but distinct enough that it does not cause disease, it becomes the ultimate weapon to fight the plague. It leads to immunity without causing death. Those first infected with cowpox are safely protected against the related smallpox. That is what a vaccine does.

Rather than think about vaccines as creative constructs of humans, another way of viewing them is as partnerships. Just as the wasp forms a mutualism with the polydnavirus to help protect its eggs, Jenner discovered that we could use cowpox to protect our children.

Although we think of vaccines as sophisticated examples of human-developed technology, the vast majority of vaccines in current use are viruses or parts of viruses. Some, like the smallpox vaccine, are simply live viral vaccines. In other words, they're just viruses we inject into people (or animals) to create an immune response that will protect against another more deadly virus. Others, like the oral polio vaccine and the measles, mumps, rubella (MMR) vaccine, are *attenuated virus* vaccines—live viruses that we have bred in the lab to make less deadly and used in effectively the same way. Some, like the influenza vaccines, are *inactivated virus* vaccines—viruses we have made incapable of reproducing themselves yet can elicit an appropriate immune response. They are still viruses. Others, like the hepatitis B vaccine and human papillomavirus (HPV) vaccine, use selected parts of the virus. The point is that pretty much the entire contemporary science of vaccinology uses viruses themselves to protect against other viruses. Safe viruses are some of the best friends we have in fighting the deadly ones.

The utility of using microbes to protect us against infectious diseases seems clear enough. But can microbes help us to control chronic diseases? The answer increasingly is yes.

Introductory courses in public health make firm distinctions between infectious and chronic diseases. They place infectious diseases like HIV, influenza, and malaria on one side of the aisle and chronic diseases like cancer, heart disease, and mental illness on the other. Yet these distinctions do not always hold up to greater scrutiny.

In 1842 Domenico Rigoni-Stern, an Italian physician, looked at the patterns of disease in his hometown of Verona. Among the things Rigoni-Stern noticed was that the rate of cervical cancer appeared to be substantially lower among nuns than married women. He also noted that behavioral factors like age at first sexual intercourse and promiscuity seemed related to the frequency of the cancer. He concluded that the cancer was caused by sex.

While sex itself did not end up being the cause of cervical cancer, Rigoni-Stern was on exactly the right track. In 1911 the young scientist F. Peyton Rous injected tissue from a chicken tumor into healthy chickens, while he was working at the Rockefeller Institute for Medical Research (now the Rockefeller University). Rous found that the injected tissue caused precisely the same type of cancer in the healthy chicken recipient. The cancer was transmissible! The virus that causes that chicken cancer—

Dr. Francis Peyton Rous, ca. 1966.
(New York Public Library / Photo Researchers, Inc.)

now called Rous sarcoma virus after its discoverer—was the first virus demonstrated to cause any cancer, and it won Rous the Nobel Prize. It would not be the last virus found to have a connection to cancer.*

In the 1970s the German physician-scientist Harald zur Hausen had a hunch about the cause of cervical cancer. Following the work of Rigoni-Stern and Rous, zur Hausen suspected it was caused by an infectious agent. Unlike the scientists of his time who thought that the cause was herpes simplex virus, zur Hausen believed that the virus that caused genital warts, the papilloma virus, was the culprit. Zur Hausen and his colleagues spent much of the late 1970s characterizing different human papillomaviruses from warts of various sorts and looking to see if they could be found in tissue samples that came from biopsies of women with cervical cancer. In the early 1980s they finally hit pay dirt. They discovered two papillomaviruses, HPV-16 and HPV-18, in a high percentage of biopsy specimens. Today, these two viruses alone are considered to account for up to 70 percent of cervical cancer.

Zur Hausen, like his predecessor Rous, received the Nobel Prize for his breakthrough. And the research they conducted went on to form the foundation for a vaccine against cervical cancer. In June 2006, Merck received approval from the US Food and Drug Administration (FDA) to market Gardasil, an HPV vaccine. Like the other vaccines discussed earlier, Gardasil uses elements of the human papillomavirus itself to elicit an immune response that prevents those inoculated from being

* Interestingly Rous didn't win the Nobel for another fifty-five years, probably the longest period of time between a key discovery and the award of the prize! His finding was not well received in the field at the time, but some scientists recognized the importance of the discovery as he was nominated to the Nobel Committee in 1926.

infected if they later have contact with the actual virus. In the case of Gardasil, the vaccine utilizes virus-like particles (VLPs) that look like the actual viruses but have no actual genetic material so they cannot replicate themselves. And the vaccine works. By preventing infection from the types of human papilloma virus that cause cervical cancer, the vaccine effectively prevents most of the deadly cancer.

Chronic diseases are notoriously difficult to treat. Whether for cancer, heart disease, or mental illness, treatments rarely return people to their pre-disease condition, and in many cases there are no treatment options at all. When a chronic disease is found to be caused by a microbe, the potential for cure and prevention improves dramatically. Cervical cancer, for example, which once required invasive, damaging, and only sporadically effective treatment, can suddenly be prevented by the deployment of a vaccine. Microbes make for low-hanging fruit when it comes to preventing and possibly curing chronic disease.

Cervical cancer is not the only chronic disease that is caused by a microbe. Liver cancer can be caused by both hepatitis B virus and hepatitis C virus. Researchers are currently exploring the possibility that prostate cancer, one of the leading causes of cancer death in American men, can be caused by xenotropic MLV related virus (XMRV). Stomach ulcers can be caused by the bacteria *Helicobacter pylori.* At least some types of lymphotropic virus, a virus family we discussed in chapter 9 and that we've discovered among the hunters we worked with in central Africa, are known to cause leukemia. It's even possible that heart disease, the culprit in one-third of US deaths and countless deaths worldwide, has an infectious component. The innovative American evolutionary biologist Paul Ewald, who has written on the connection between infectious agents and chronic disease, suggests that the interplay between *Chlamydia pneumoniae* and environmental factors may be to blame for heart attacks, strokes, and other cardiovascular illness.

In some cases viral causes are suspected but have not yet been confirmed—perfect fodder for eager scientists. The distribution of type I diabetes cases suggest a possible connection with an infectious agent, but none to date has been identified. My own research team and our collaborators recently began work on a grant from the National Cancer Institute to screen tumor specimens from multiple types of cancer in search of viruses. It's exploratory research, but the potential benefits as we find them could be monumental.

Even some mental illnesses may result from infections with microbes. As we've seen, microbes can have an impact on behavior. Toxoplasma alters very specific neural circuits in rodent brains to decrease their fear of cats and thereby increase the chances that the parasite can complete its life cycle by ending up in a hungry cat. Rabies causes fear of water and increases aggressiveness in those infected with it, which helps accumulate virus in saliva and deliver it through a potentially fatal bite.

With these prominent examples of behavioral manipulation, it's an obvious leap to suspect that microbes could play a contributing role in mental illness, a subject that has been the focus of a researcher at Johns Hopkins Medical School for some years. Robert Yolken studies a range of disorders, including bipolar disorder, autism, and schizophrenia, examining them closely to see if microbes might play a role. His primary focus is schizophrenia.

Schizophrenia certainly seems to invite discussion on links with infectious agents. For years, researchers have noted a relationship between seasonality of birth and schizophrenia: children born in winter months are more likely to develop schizophrenia than those who are not. This finding has long been thought to suggest that wintertime illnesses such as influenza, infecting either the pregnant mother or infant, may predispose an individual toward schizophrenia, although the results remain unclear for now.

Yolken's most recent focus has been *Toxoplasma gondii*, or simply toxoplasma. He and others in the field have put together a plausible if perhaps not fully definitive case for the parasite's role in this devastating mental illness.* Multiple studies have found a correlation between schizophrenia and the presence of antibodies to toxoplasma. Some adults who experience the onset of toxoplasma disease experience psychological side effects. And antipsychotic drugs used to treat schizophrenia have also been seen to have an effect on toxoplasma in laboratory cell cultures. In a sign of the intense research that has surrounded the subject of schizophrenia, studies have documented that individuals with schizophrenia have had more exposure to cats than unaffected controls. Together these and other studies point to a connection. This connection still faces challenges since the parasite is not likely to be involved in all cases of schizophrenia, a disease that also has important genetic determinants.

A virus may also be the cause of a complex, controversial, and somewhat mysterious disorder. Chronic fatigue syndrome (CFS) is a debilitating illness with no known origins and a variety of nonspecific symptoms: weakness, extreme fatigue, muscle pain, headaches, and difficulty concentrating, among others. Most people who have stayed up all night studying for a final or pushed themselves too hard at the gym will recognize these

---

* Interestingly, *Toxoplasma gondii* may provide a scientific explanation for a commonly stereotyped set of behaviors. Recent attention to the "crazy cat lady syndrome," as it is referenced in an article in the *New York Times*, points out that cat-hoarding behavior resembles the behavior of rodents infected with toxoplasma—affinity toward cats and immunity to the smell of their urine, for example. To date there have not been scientific studies to prove or disprove this hypothesis.

Dr. Robert Yolken with one of his subjects. (*McClatchy-Tribune / Getty Images*)

symptoms as familiar and common. They are also common symptoms for many other medical conditions, making it difficult to eliminate other possible root causes. As a result, medical experts and members of the public have debated the authenticity of CFS as a unique disorder. However, recent studies support those who argue that CFS is a genuine disease. Following several studies with contradicting results, a study published in August 2010 found a correlation between CFS and a virus in the murine leukemia virus family. More research is necessary to establish a causal link between MLV and CFS, but the finding has offered hope to many.

As with cancer, a microbial cause of schizophrenia or CFS would invite quick and possibly important new diagnostics, therapies, and vaccines for these chronic disorders, which cause great pain and discomfort to victims and families. In the case of cervical cancer, the vast majority of the illness is ascribable to human papilloma virus, so a vaccine preventing it could be developed. This is not always the case. If only a percentage of people who suffer from schizophrenia or CFS do so because of a virus, it will make the associations more complicated and the discovery of links more challenging. Yet it's worth the effort. Many chronic diseases lack good treatment options, and our ability to create vaccines and drugs for microbes is legendary. Wouldn't you want to vaccinate yourself or your children for schizophrenia or heart disease? Even if it only protected them from one of a handful of causes of the illnesses? One day, we hope, you will be able to do just that.

Using one microbe to prevent another microbe from causing disease is pretty amazing. But how about using a microbe to actually address the disease directly? This is something that's increasingly explored in the nascent field of *virotherapy*.

All viruses infect cells as part of their life cycle, and they don't infect cells randomly. As we've discussed, viruses infect cells in a lock-and-key manner: they enter into those cells that have particular proteins, or cell receptors, on their cell surfaces that the virus recognizes. If a virus existed that recognized and infected only cancerous cells, for example, then the virus could theoretically burn through those cells, killing the cancers along the way. The hope, of course, would be that when they were done with the cancer cells, they'd have nothing to infect and would die off.

Just such a virus exists. The Seneca Valley virus is a naturally occurring virus that appears to specifically target tumor cells living at the interface of the nervous and endocrine systems. It reproduces in the tumor cells, causing lysis, or rupturing and death of the cells. When released, it spreads to new tumor cells to continue its work. Now that's a gentle virus!

Seneca Valley virus was discovered in a biotech company laboratory in Pennsylvania's Seneca Valley. The virus had likely contaminated cell cultures from cattle or pig products commonly used in the laboratory. It was isolated and found to be a new virus in the picornavirus family, which includes polio. Testing showed that the virus had amazing selectivity to cancerous cells in the neuroendocrine system yet failed to infect healthy cells. This is a good reminder that not all viruses that cross the species barrier do harm.

The Seneca Valley virus is not alone. The small but growing group of virotherapy researchers use and adapt a range of viruses, including herpes virus, adenovirus (one of the viruses that causes colds), and the measles virus—to create viral therapies that can knock down cancer. Probably the most advanced among them is a herpes virus therapy developed by a biotech firm called BioVex, which is in the last stage of trials to determine its ability to control head and neck cancer. While the results of the trial have not yet been released, Amgen, a Fortune 500 biotech company, recently entered into the final stages of a deal to acquire the smaller BioVex as well as its herpes virus therapy.

What about viruses that interfere with other viruses?

One brilliant example is a wonderful little virus called GB virus C, which appeared in chapter 5 and is found in a high percentage of people. This odd-sounding virus is in the same family as hepatitis C virus, but it certainly doesn't kill us. In fact, it can save us.

In an incredible study published in the top medical journal the *New England Journal of Medicine* in 2004, researchers showed that infection with the GB virus C could help prolong the lives of men who were infected with HIV. When examined five to six years after infection with HIV, men without detectable GB virus C were nearly three times more likely to die than those who had active GB virus C infections. How GB virus C acts to save AIDS patients is still unclear, but it appears that it might interfere directly with HIV. Whatever the mechanism, this tiny organism has likely prolonged millions of lives during the course of the current pandemic.

Viruses can also interfere with other kinds of microbes—bacteria can get sick too. Viruses infect all forms of cellular life, whether bacteria, parasite, or mammal. As we discussed in chapter 1, while nonspecialists tend to see microbes as a homogenous bunch, nothing could be further from the truth. All of the cell-based life forms (bacteria, parasites, fungi, animals, plants, and so forth) are thought to be more closely related to each other than they are to viruses.* Furthermore, parasites fall into the class of life called eukaryotes and are more closely related to us than either they or we are to bacteria.

A fascinating Harvard virologist now at the Texas Biomedical Research Institute, Jean Patterson became interested in just this phenomenon in the mid-1980s. While her main focus had been viruses, she wanted to look closer at a group of parasites called protozoa, which includes malaria and leishmania, a harmful protozoan parasite transmitted to humans by the bite of the sand fly. Patterson was interested in how the parasites translated their genetic information into action, and she became fixated on discovering a virus that could infect this interesting parasite.

In 1988 Patterson and her colleagues discovered a small virus that naturally infects leishmania parasites; they were the first to characterize a virus from this group of parasites. Viruses

* There is still debate on the relationship between viruses and cell-based life. In fact, viruses may not even all be related to each other. Some may have originated as the DNA of cell-based life forms while others may be descendants of life forms that predated the emergence of cell-based life.

that infect parasites could provide natural systems for parasite virotherapy. And as with the cancer-killing viruses, parasite viruses could potentially be adapted for efficiency and safety.

I've personally spent a reasonable portion of my professional life studying protozoa parasites. First, as a doctoral student working in Malaysian Borneo with my veterinary colleagues Billy Karesh, Annelisa Kilbourn, and Edwin Bosi, we tried to understand malaria in wild and captive orangutans.* More recently, my colleagues and I searched for the origin of malaria in central Africa, a subject discussed in detail in chapter 3. Could it be possible that in some of our vials holding an ape malaria parasite resides a new malaria-infecting virus? One that could potentially kill our own deadly malaria, *Plasmodium falciparum*?

When most people think about microbes, they frame it as a battle of people versus bugs. Perhaps if they're being a bit more creative, they'll consider the battles among the microbes themselves. But the reality is even more interesting than that. We're part of an incredibly rich community of interacting microbes—with hugely complicated collaborations, battles, and wars of attrition with each other and ourselves.

---

* My doctoral research was conducted largely in Borneo, in the Malaysian state of Sabah. I was lucky enough to get the assistance of the world's most prominent wildlife veterinarian, Billy Karesh, who at the time was at the Wildlife Conservation Society. Billy took me under his wing, let me take part in his project, and introduced me to his Malaysian colleagues. I watched them conduct absolutely amazing work, tranquilizing wild orangutans with dart guns and moving them from small disappearing forest fragments to a large reserve that the Malaysian government had set aside for conservation. From my own research perspective, I had the invaluable opportunity to get specimens from elusive wild orangutans while they were being transported! During my time there, I spent many months working on a daily basis with Annelisa Kilbourn, the extraordinary wildlife veterinarian who died tragically during a plane crash some years later while working with gorillas in central Africa.

Consider the human body. Only about one out of every ten cells between your hat and shoes is human—the other nine belong to the masses of bacteria that coat our skin, live in our guts, and thrive in our mouths. When we consider the diversity of genetic information on board, only one out of every thousand bits of genetic information on and in us can properly be called human. The bacteria and viruses represented by thousands of species will outnumber the human genes every time.

The sum total of bacteria, viruses, and other microbes present in our body is called the microbiota, and the sum total of their genetic information is called the microbiome. A new science has developed in the past five years to characterize the human microbiome. Empowered by new molecular techniques that bypass the nearly impossible task of individually culturing each of the thousands of microbes, scientists are rapidly figuring out exactly what the overall community of human and microbial cells in our bodies consists of.

The findings coming out are fascinating. Our guts are teeming with a complex assemblage of microbes, many of whom are long-term residents. They are not simply free riders. A great deal of the plant material we consume requires bacteria and their enzymes for digestion; human enzymes alone would not do the trick. And how the community of microbes is structured makes a big difference.

In a pivotal series of studies, Jeff Gordon and his students and postdocs (many of whom are now successful professors themselves) showed just how important the communities of bugs in our guts actually are. They have demonstrated that obesity is associated with a decreased relative abundance of one particular group of bacteria—the Bacteroidetes.

In another elegant study, Gordon and his team showed that the obese microbiota increases the amount of energy that can be obtained from food. In the final coup de grâce, they showed that altering the gut microbiota of normal mice with the obese microbiota results in significant weight gain. Very simply, bacteria in our guts play a role in obesity. Just as we saw with cervical cancer, a microbial cause of a chronic disease may point to an easier method to solve it. One day we may very well use a combination of probiotics and antibiotics to subtly alter our gut microbiota and to help us maintain a healthy weight.

Perhaps not surprisingly, the teeming masses of microbes in our guts also play a role in how we're affected by deadly microbes. In the case of salmonella, a deadly bacteria and one of the leading causes of food-borne illness, it's been known for some time that the biggest risk factors for the disease are eating eggs away from home and using antibiotics. Eating eggs is a risk factor since chickens infected with the bacteria can contaminate them. The antibiotic use, however, has long presented a mystery.

Recent research on gut microbiomes may shed some light. Justin Sonnenburg, a Stanford professor, is conducting important work to do just this. He uses an incredible system for maintaining germfree mice in a laboratory. The rodents live in completely sterile conditions—even to the point where their food is autoclaved before they eat it, eliminating any potential microbial contaminants. The germfree rodents provide a perfect model for picking through the exact determinants of different gut microbiota on the conditions of their hosts.

While it's long been suspected that antibiotic use kills helpful microbes, thus damaging the natural shield that our gut microbes provide against new and invasive bugs like salmonella, it's still not clear exactly how this happens. In the future, the work done in Sonnenburg's lab should tell us.

There are gentle microbes out there—bugs that help us, defend us, and live quietly within us doing no harm at all. If we could accurately determine which of the microbes on our bodies and in the environment were beneficial to us and which were rogue, we'd find something pleasantly surprising: the harmful ones are certainly in the minority. The goal of public health should not be to have a completely sterile world but to find the rogue elements and control them. A key part of addressing the nasty microbes will be to cultivate the microbes that help us. One day soon, the way we protect ourselves may be by propping up the bugs that live within us rather than knocking them down.

►12◄

THE LAST PLAGUE

The large, brightly lit, white-walled room appears at once chaotic and oddly organized. Young kids in their Silicon Valley uniforms of hoodies and sneakers sit hunched over laptops, talking on the phone and instant messaging while simultaneously mashing together and analyzing massive amounts of disparate data. Large monitors with maps and streaming news line the walls. There are no windows, so it's hard to determine if it's daytime or evening. Discarded coffee cups and junk food wrappers also fail to reveal the hour. Occasionally an older group wearing suits and formal business attire enters, chats, and then just as quickly disappears. As the discussion comes into focus, its purpose emerges: a 24-hour global situation room for emerging diseases.

At the top of the agenda of this California control room are Nigeria, Dubai, and Suriname. Clear signals from the masses of data collected have elevated their risk profile to "regular alert," which means that roughly 20 percent of the team's effort is focused on getting more data, interfacing with on-the-ground team members, and conversing with local and international health leaders. In the case of Suriname, the problem has already hit the news. Hospital admissions have been up over the past twenty hours, and a local newspaper article has hit with the potentially dubious report of cholera. In Nigeria and Dubai, the events so closely tracked in this room haven't yet gone public. But they will.

A closer examination of one of the young analysts reveals what kind of data she's crunching—disease data. With three active computer screens, she's following the frequency of "chief medical complaints" filtered and forwarded from an early but robust cell-phone–based electronic medical record system located in Lagos. The frequency with which the users are reporting severe fever has increased steadily from baseline over the past thirty hours, and it matches independent over-the-counter drug purchase data for medications treating fever and malaise. The Twitter and Google trends on terms related to acute viral illness also seem to match. People are "telling" them they're getting sick. The analyst's teammates at the central African headquarters in Yaoundé have been on the phone with clinics for hours. The lab results are still pouring in, but the uptick is due to none of the usual suspects; it's not malaria or typhoid, nor is it Marburg or Ebola.

On another screen, our analyst Skypes with someone on the hacker team. They're opening the repository data feed with the lab in Lagos. Soon the local group will have the potential to upload reams of new genetic data from specimens that are just now being examined. Computer algorithms and bioinformatics engineers will search for the needle in the haystack—the new virus that appears to be killing people in west Africa.

Our analyst's immediate boss is the Room Controller—a specialist who must weigh the data and evaluate the rankings that the computer systems are providing. Does Nigeria move up to "full alert" status as the algorithms recommend? How does this compare with the faint but potentially frightening bioterror chatter coming out of Dubai, where aggregated purchase data suggests that someone is buying equipment normally used to grow huge batches of bacteria? And how will it measure during the daily briefing against the longer-term trends examined in the chronic-infection group that seeks to identify unseen and creeping killers, rather than the more immediate but obvious flare-ups of the control room? What happens in this room is fast paced, globally interconnected, and potentially world saving.

This scenario is fiction. There is no such control room—yet. The reams of data from electronic medical records in Lagos do not yet exist, and the data from pharmacies are not yet well coordinated and compiled. But while we're not there yet, the control room is exactly what we need—an innovative group devoted entirely to understanding and analyzing biological threats and catching them before they become disasters.

I've spent time in the closest equivalents our planet currently has to such a control room. During the beginnings of the H1N1 influenza pandemic, I visited with Scott Dowell—the head of the CDC's Division of Global Disease Detection and Emergency Response—in the CDC's control room, where the team rapidly responded to the mounting reports of illness in Mexico. I've also spent time in the WHO's control room used during pandemic and other health emergencies. My organization, Global Viral Forecasting, is part of the WHO's Global Outbreak Alert and Response Network. Sadly, bureaucracy, insufficient and ever-shifting funding, and constantly changing objectives from higher up the food chain hamper both the CDC and the WHO. These organizations need to grow stronger and better equipped, and they desperately need more funding. But even then, more will be needed.

In this final chapter, my hope is to review where we've come so far in the book. How do the history and advances stack up—in our favor or against us?

I'll also try to answer some questions that I often get asked as a virologist. What steps do I personally take to mitigate my risk of infection? How should individuals evaluate a pandemic or biological threat while it's occurring?

I'll also do my best to answer the broader questions about what I believe is needed for the planet. What are the largest impediments to controlling future pandemics? What are we doing to get to the vision of the futuristic control room imagined above?

In the previous chapters, I've tried to provide a picture of where we currently stand in regard to pandemics and other microbial risks—to explain the central characters, the microbes, on their own terms and to examine how the major events in our history have affected our relationship with them.

What we've seen is that a range of early events created the conditions for a perfect viral storm. The advent of hunting in our biological lineage created a species that suddenly interfaced with animals, permitting a flood of new microbes into prehumans. And the near-extinction event we experienced likely left us ill prepared to deal with them.

We saw how an increasingly populous and interconnected world served to push us toward the center of the storm. Domestication of animal populations, the growth of urbanization, and our miraculous transport system tied together populations in ways unprecedented in the history of life on our planet. Particular human flourishes, including transplantation and injection, provided completely new routes for disease agents—whether natural or purposefully introduced—to spread and create havoc.

Chapters 9 and 10 gave a sense of the contemporary tools we have that might allow us to countervail against the rising threat of pandemics—new techniques for diagnosing microbes and new methods for monitoring people and communities. Contrasting with most of the book's presentation of harmful bugs, chapter 11 explored the emerging uses and benefits of many harmless microbes.

• • •

As a professional microbiologist, one common question I get is how do I personally mitigate my risk of infection? For starters, I always keep my vaccines fastidiously up-to-date. When I am in malarious regions, I take malaria prophylaxis religiously. I didn't always do that, but I learned the hard way how important it is.

During winter months, I'm aware of transmission routes for respiratory illnesses, and I do my best to decrease my risk of acquiring them. Public transport is a notorious risk due to the mass movements of humans through it, so I try to wash my hands or use a simple alcohol-based hand sanitizer when leaving subways or planes. Likewise, I'm aware of times when I'm shaking large numbers of hands and try to wash them soon after or avoid touching my nose or mouth unless necessary. Certainly doing your best to ensure consumption of clean food and water is important, and so is working to limit risk associated with unsafe sex. Of course, these answers really depend on who you are and where you live. Sadly, access to clean water, vaccines, good malaria drugs, and condoms are by no means universal—but they need to be for *everyone's* sake.

Perhaps of equal interest is how to evaluate the news reports and assess risk when an outbreak occurs. This can be done by focusing on a few particular features of the epidemic. How is the microbe spreading? How effectively is it being transmitted? What percentage of people that it infects is it killing? A very deadly epidemic that *doesn't* seem to be spreading is less worrying than a nominally deadly pandemic that's moving at a fast and efficient clip. Things that seem terrifying, like the Ebola virus, aren't always global risks. And things that seem benign, like HPV, sometimes can be devastating. Fortunately, basic facts about transmissibility and deadliness can help anyone evaluate the risk.

Assuming that living in one location or courting a certain quality of life makes you immune from the risk of a pandemic is wrong. While HIV didn't spread around the world randomly affecting people, it affected very poor people and very wealthy people alike. It affected people with almost no access to health care and, in the case of hemophiliacs, some of the people with the best health care in the world. We're all on one interconnected planet.

Among the most pressing questions I get when I speak to audiences globally about these issues is "Ok, I get it and now I'm scared. How do we deal with this?" One of the greatest impediments to predicting and preventing future pandemics is the notion that pandemics occur randomly and are inherently neither predictable nor preventable. If I have done nothing else in this book, I hope I have repudiated these ideas. Prediction and prevention of pandemics will not be easy, but there is much we can do right now, and the advances that are steadily occurring will allow us to do even more in the future.

Not having an active public health mind-set of pandemic prevention has led to hugely inefficient systems in the past. Among the best terms I've heard to describe these inefficiencies and overreactions is *disease du jour.* When we have an influenza threat, we drop everything and focus on mitigating risk from future influenza pandemics. When we have SARS, we focus on unknown respiratory diseases. The list goes on.

One day we may be able to rank the greatest future risks for pandemics, but for now we cannot. We know that they'll almost certainly be microbes that come from animals and that some spots around the world pose greater risks for their entry. We need resilient systems that don't assume the next threat will be influenza or SARS or whatever the au courant infectious disease happens to be. The systems should be generic and forward focused. They should target the unknowns and the general patterns that gave us our past pandemics, rather than any of the specific pandemics we've had. This doesn't mean we should disregard the excellent global influenza surveillance systems or the wonderful work done by my colleagues, like Derek Smith, who uses data on global samples of seasonal influenza to predict the next year's strain and develop vaccines against it. But we should also recognize that those systems will help us mitigate future influenza risks—*not* the risks associated with the next unknown agent.

The good news is that the years of harping on about pandemic prevention have begun to pay off. Under the leadership of the forward-thinking government official Dennis Carroll, who directs the Avian Influenza and Other Emerging Threats Unit at the US Agency for International Development (USAID), a large program to understand and develop global capacities to counter Emerging Pandemic Threats (EPT) has flourished, a program that I am proud to be a part of. Other organizations like Google.org and the Skoll Global Threats Fund have identified prediction and prevention of pandemics as a central goal and brought fascinating technological and entrepreneurial perspectives to the problem.

The US Department of Defense (DoD) has also played a pivotal role. While the press continues to focus almost exclusively on their involvement in wars, the truth is that their international disease tracking and control systems are among some of the strongest in the world. In the name of protecting global troops and combating biological threats, organizations within the DoD like the Defense Threat Reduction Agency (DTRA) and the Armed Forces Health Surveillance Center (AFHSC) have devoted serious technical expertise and resources to understanding the nature of threats, finding diagnostics and treatments, and developing local capacity around the world to help engage microbiologists to fight natural pandemics.

I'm lucky to be able to work with all of these groups. Together with a number of other organizations, we're beginning to craft the strategies needed to seed a full paradigm shift from response to prevention in the area of global pandemics. The hope is that it will not take as long as it took public health officials to embrace prevention of diseases like heart disease and cancer, but however long it takes it is imperative we move in this direction.

Another huge problem that interferes with our ability to stop future pandemics is the inaccurate assessment of risk by the public. This is something that I heard referred to as "risk literacy" at the 2010 Skoll World Forum by one of the early founders and constant supporters of the field of pandemic prevention, Larry Brilliant. Larry, who won a prestigious TED prize for his wish to "help stop the next pandemic" has been a central leader in getting this movement off the ground through leadership at Google.org and now the Skoll Global Threats Fund. Larry was a key team member on the smallpox eradication program, so he couldn't be better suited for the task.

The term *risk literacy* is an important one. The idea is that part of the solution is having an informed public that can understand and appropriately interpret information on pandemics.

Risk literacy, the ability to distinguish between different levels of risk severity, is not only for policy makers. Effective response to natural disasters depends on individual people and how well they stay calm and follow instructions. The constant barrage of threats articulated by the media has led to chronic risk habituation. The only way to break that logjam is for everyone to understand risk, to be able to assess the differences between different kinds of disasters, and to respond appropriately to them.

Widespread risk literacy will help the public support the massive government expenditures that will be needed to appropriately predict and prevent pandemics. It will give us a sense of how best to expend funds. From April 2001 to August 2002, a period which includes the 9/11 attacks, it's estimated that around eight thousand people in the world died from terrorism. From April 2009 to August 2010, the same period of time but eight years later, over eighteen thousand people were confirmed dead from the H1N1 pandemic alone—a pandemic dismissed by the public as insignificant. And that number is certainly an underestimate. I'm not claiming that proportionality in deaths is the only factor we should take into account when preparing for threats. But the trillions of dollars spent to prevent terrorism seem wildly disproportionate to actual risks when we put the threats in their proper context.

Perhaps my favorite question to answer is "What are you doing about all of this?" During the past two years, I've had the honor of leading an incredible group of scientists and logisticians who work around the clock and around the world to develop and deploy systems with a single objective—to catch pandemics before they spread and stop them.

This work has come out of the past fifteen years of research I've done, much of it described in these pages. In 2008 I made a decision, seen by most of my colleagues as lunatic at the time, to quit my job at UCLA as one of the increasingly rare lifelong tenured professors. I left to start Global Viral Forecasting (GVF), an independent organization devoted to monitoring global medical intelligence and using it to catch pandemics early.

Based in San Francisco, where I continue to teach at Stanford, GVF uses every possible tool at its disposal to identify and defuse epidemics. In government or academia, there's always a sense that you should stick to a particular approach to solving a problem. Microbiologists use microbiology and epidemiologists use epidemiology. At GVF the tool is unimportant—we care only about the goal of early actionable intelligence on the trends and movements of infectious diseases in human and animal populations.

We combine on-the-ground epidemiological tools for early detection of outbreaks and documenting the microbes in human and animal communities with cutting-edge information and

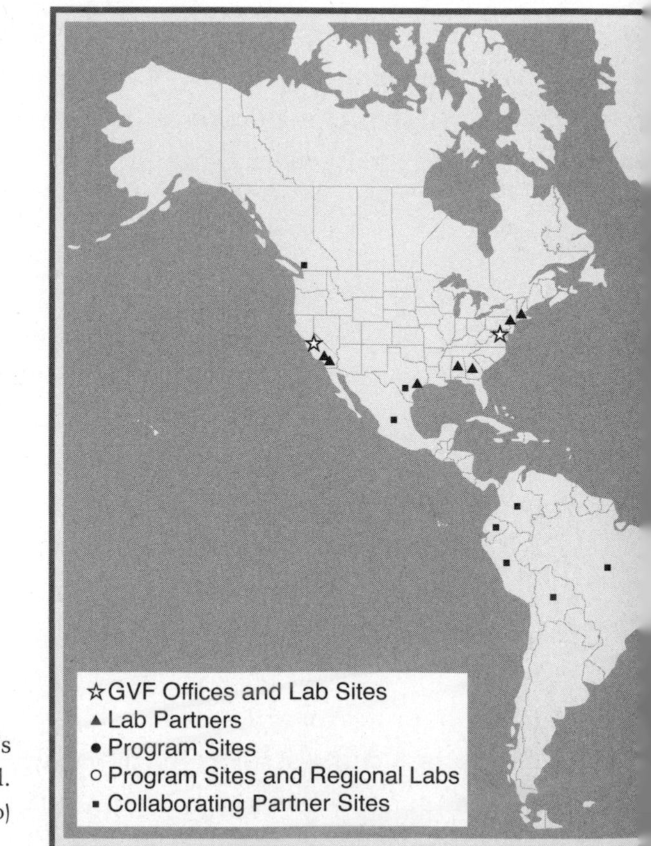

Global Viral Forecasting's sites around the world. (Dusty Deyo)

communication technologies to monitor "digital signals" of the next plagues. The goal of the organization (to chart infectious diseases and to reach a point when we can predict and even prevent pandemics) is ambitious, but our laserlike focus on a single objective also provides a certain freedom—if it's not on task, we shouldn't be doing it.

No matter what novel technologies and tools we use, there is nothing like on-the-ground information. So the backbone of our work is ongoing field efforts in countries around the world. The objective is to understand the microbes present in animals that might jump into humans. We also chart and follow the microbes that are already in humans that might cause diseases

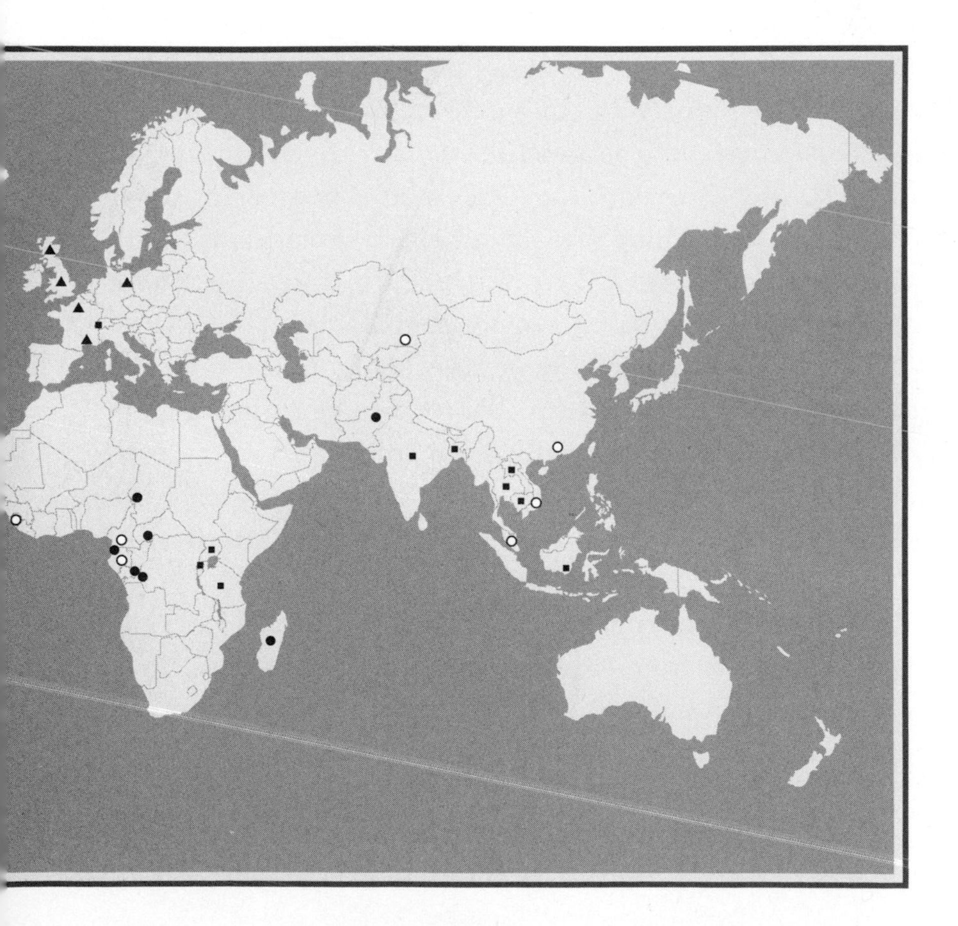

in ways we haven't identified yet. And finally, we try to catch new outbreaks and epidemics early, before they hit the radar of traditional health and media organizations.

To do this we conduct regular surveillance in hospital and clinical settings. We also focus energy on particular groups of people we believe to be *sentinels*—populations that because of their locations or behaviors will, like the canaries in the coal mine, become infected first before a microbe enters wide circulation. Through ongoing monitoring of hunters, we've found a range of new microbes that had not previously been seen. Using this kind of painstakingly collected surveillance data, we've also documented evidence that known viruses, like human parvovirus 4, have spread much more substantially than was previously considered possible.

Our model for studying sentinel populations that represent portals of entry for new animal microbes on their route to pandemics has proved very successful. Along with our partners in the USAID-EPT Program, the DoD, and a number of other partners, we are pushing that model out to over twenty countries around the world. Yet more is needed. In each of those countries, we need to monitor more people who are at risk of acquiring new and potentially transmissible agents from the animals they have contact with. We need to be active in more regions of each country in which we work, and we need to expand the effort to many more countries. In many ways, the work of surveillance for potential pandemics has only just begun.

In addition to studying the points at which agents enter the human population from animals, we also work extensively to monitor critical populations that are central in the networks wherein microbes spread. For example, we carefully follow groups of individuals who receive regular blood transfusions. Since some of these individuals will receive many hundreds of transfusions from numerous people, they will be infected early and can help indicate when something new is on the move. There are many groups of people who are at the center of hubs and will likely be among the first to get a new agent that's spreading—health care workers and airline flight attendants, to name just two. Critical to our work is to continue bringing more of these populations into our monitoring system.

Animals are vital too. In chapter 9, I outlined the way that the head of GVF's Ecology Team, Mat LeBreton, and I developed an approach to using simple pieces of standard laboratory filter paper to quickly amass large numbers of blood specimens from animals in the places where we work. To that we have now added monitoring of so-called animal die-offs. Each day, somewhere on the planet, a group of wild animals dies just as the apes in Cameroon who succumbed to anthrax did. Small animal outbreaks provide unique opportunities to understand what microbes are out there.

Animal die-offs can also presage a human outbreak, as occurs in the case of yellow fever, for example, in South America. Often after the deaths of forest monkeys, human settlements will become infected with the deadly virus. Yet currently very few animal die-offs are identified. Now with help from some of the hunters we work with in biodiverse forests throughout the world, we are beginning to set up a system to investigate more of them. Ideally, we should know whenever a group of animals dies anywhere in the world, but in our current reality we have almost none of this critical information.

While the majority of our fieldwork at GVF is catching any one of a vast range of new agents, in some of our sites we focus largely on a single known agent. In Sierra Leone, Joseph Fair, the virologist and field epidemiologist who leads GVF's field and laboratory efforts, has conducted challenging, cutting-edge research to understand and control a deadly virus called Lassa fever. Lassa is a fascinating and dangerous virus that passes from rodents to humans when it comes into homes and contaminates food.

Lassa causes the same sort of devastating symptoms as Ebola and Marburg. The Lassa site that Joseph has developed in west Africa provides a model that allows us to understand these agents and how best to anticipate and respond to them. Other than Lassa fever, all of the hemorrhagic fever viruses—the deadly class that includes Ebola and Marburg—occur sporadically. Lassa on the other hand, is a regular part of life in these regions. Because it is nearly impossible to really monitor viruses that occur only occasionally, Joseph and his colleagues have used the important sites in Sierra Leone to study and learn the best ways to catch and control these kinds of viruses before they spread. For the fans of outbreak movies, it is a sexy site with a high level of biological containment and workers scrambling to save lives while risking their own. Yet its importance goes far beyond that. If we learn to anticipate and respond well to Lassa, it will help us to make sure that viruses like Ebola and Marburg never get out of control.

Perhaps one of the most exciting things we do at GVF has nothing to do with cutting-edge science. It's prevention.

Among the most substantive risks for the emergence of novel pandemics is the close contact between humans and animals, particularly wild mammals. Changing human behavior to decrease this sort of contact can begin long before we have the ideal prediction systems in place.

Karen Saylors, a devoted medical anthropologist who has been on our team for years, living in field sites in central Africa and now in our headquarters in San Francisco, spends long hours working with GVF teams and other colleagues in field sites around the world to prevent new pandemics. We have conducted this work for years but are now pushing to expand quickly. For nearly a decade Mat LeBreton and his colleague Joseph LeDoux Diffo and others have developed a program we consider central to our work—the Healthy Hunter Program. In this program we try to find ways to decrease the risks of acquiring new viruses in the wild-game hunters we work with in central Africa. Since this is the place and manner in which HIV emerged, we consider the work vital. But it's not easy.

I can remember in the early days of my work in central Africa the response I got from hunters when I described my sense of the risks associated with their hunting and butchering wild mammals: "We have done this for years and our parents and grandparents did this for years. Surely it cannot be as much of a risk to us as the many other things that kill us here." It was a response I was to get in all the places we worked. And there is truth to it. In an environment where malaria, unsafe water, and poor nutrition kill on a daily basis, the risk of acquiring a novel agent from an animal seems minimal. And in some ways it is.

The problem is a perfect example of a tragedy of the commons. For the vast majority of individual subsistence hunters, the risk of acquiring a novel deadly disease will be less than the nutritional and other costs of not hunting. Yet when many thousands of people in regions with high diversity of microbes hunt wild animals, we enter into a situation that can result in the emergence of some new agents—agents that could devastate the entire planet. The problem is not the hunters' alone. It belongs to all of us.

While we work diligently to help inform these people of the risks associated with hunting, we also recognize that the real enemy is rural poverty. To solve this universal problem, we need to do more than explain risks. We need to devote ourselves to helping to find viable solutions to the nutritional needs of rural populations. We need to help them find alternatives to unsafe hunting, and we cannot blame them for trying to feed their families. As we expand our Healthy Hunters Program to more sites, we simultaneously work with development and food organizations to provide real solutions.

If we could snap our fingers and eliminate the hunting of wild game for subsistence in viral hot spots like central Africa, southeast Asia, and the Amazon Basin, we certainly would do just that. In addition to the risks of pandemics, these practices have well-known negative implications for the biological heritage of our planet and for the food security of vulnerable populations living off non-renewable animal protein sources. Yet the solution will require real energy and resources on a global scale. It will be energy well spent. In addition to the self-serving goals of wealthy populations around the world to stop plagues and preserve biodiversity, it would also help some of the poorest populations in the world to live a reasonable life. The problem of bushmeat is not a boutique issue for those wanting to save some charismatic endangered species. It affects global health, and we cannot afford to ignore it.

As GVF looks for more partners and more resources to help extend our first efforts at changing the behaviors that allow new agents to enter into our species, we recognize there is more we can do now to prevent the activities that lead to pandemics. And some of the things we can do align perfectly with other public health initiatives. As we discussed in chapter 8, immunosuppression that occurs with AIDS facilitates the entry of new microbes into human populations. We must work to guarantee efforts to extend the antiretroviral drugs that control AIDS to even the most rural populations that have contact with wild animals through hunting. We have worked with some of the pioneers in this field—scientists like Debbi Birx, who left a successful career overseeing a productive research group at WRAIR

(Walter Reed Army Institute of Research) to lead the CDC's Global AIDS Program, which focuses on the nuts and bolts of getting antiretroviral therapy to some of the neediest parts of the world. This will help us all.

There are ways that each of us can help this process. It is vital that we all put pressure on policy makers and politicians to support long-term approaches to pandemic prevention. An informed public must push governments to provide more funding aimed at generic approaches to controlling future pandemics rather than simply focusing on a single threat.

In an ideal world we might embrace changes suggested by some in the wake of recent pandemics. At the 2009 TED conference in Long Beach, Fred Goldring, an influential entertainment lawyer, suggested that we should advocate a "safe shake," where we shake by touching elbows rather than hands. Certainly this would help to decrease the spread of some infectious agents in the same way that sneezing into an elbow rather than a hand does. To my knowledge no one has conducted detailed studies of the health impact of bowing (rather than shaking hands) in countries like Japan, but it would be expected that it should decrease the transmission of some infections. Similarly the practice of wearing surgical masks in public when ill, seen in Japan, could well dampen some bugs from spreading. Changing habits like these is incredibly difficult, but the models show that useful possibilities exist.

You may ask yourself when we will see the ideal control room that began this chapter. Though that scenario was fiction, there is no reason that we need to wait centuries or even decades for it to occur. In fact, one of our goals at GVF is to make this a reality. Our data team, headed up by Lucky Gunasekara, consists of a completely new breed of digitally minded young scientists who meld the work we do in the field with the entirely new set of data we discussed in chapter 10. Detailed data from the field and lab will soon combine with data from cell phones, social media, and other sources to create the ultimate outbreak data mash-up.

A decade ago the main structures that organized the world's information were governmental, like the Library of Congress. Yet this was not the final answer. Today, organizations like Google have used innovative methods and incentives to build tools for accessing information that we could barely have dreamt of a few decades ago. We must be open to innovation of this sort in the area of global health. It is often said that organizations like Google have helped create a *global nervous system*. If we are ever to have the equivalent of a *global immune system*, we will need to develop new approaches that combine governmental and nongovernmental systems and use the latest approaches and technology.

In fact, this has already begun. In the coming years, whether you are a head of state wary of political and economic costs of a disease catastrophe, a CEO concerned by supply-chain and staff disruption associated with the next pandemic, or a citizen worried about your family, you will have access to better, more accurate, and rapidly available data on actual outbreaks. And not just from governments but from organizations like my own, which will combine lab results in far-flung viral listening posts with international news feeds, text messages, social networks, and search patterns to create a new form of epidemic intelligence.

We live in a world fraught with risk from new pandemics. Fortunately, we also now live in an era with the tools to build a global immune system. This huge but simple idea is that we should and can be doing a much better job to predict and prevent pandemics. But the really bold idea is that we could reach a point where we become so good at this that we mark the "last plague"—a time when we are so capable of catching and stopping pandemics that we won't even need the word.

SOURCES

This section provides sources I used either directly to obtain figures or facts or indirectly as background. It also includes books (marked with an *) for those interested in further reading on topics raised in the individual chapters.

INTRODUCTION

Balfour, F. "A Young Life Ended by Avian Flu." Businessweek.com, February 3, 2004.

*Barry, J. M. *The Great Influenza: The Epic Story of the Deadliest Plague in History.* New York: Viking, 2004.

"Bird Flu Claims First Thai Victim—January 26, 2004." CNN.com World, January 26, 2004.

Centers for Disease Control and Prevention. "Childhood Influenza-Vaccination Coverage—United States, 2002–3 Influenza Season." *MMWR* 53 (2004): 863–66.

Chokephaibulkit, K., M. Uiprasertkul, P. Puthavathana, P. Chearskul, P. Auewarakul, S. F. Dowell, and N. Vanprapar. "A Child With Avian Influenza A (H5N1) Infection." *Pediatric Infectious Disease Journal* 24, no. 2 (2005): 162–66.

Clayton, D. H., and N. Wolfe. "The Adaptive Significance of Self-

Medication." *Trends in Evolution and Ecology* 8 (1993):60–63; doi: 10.1016/0169-5347(93)90160-Q.

"Cumulative Number of Confirmed Human Cases of Avian Influenza A/(H5N1) Reported to WHO." Global Alert and Response (GAR), World Health Organization, December 9, 2010; www.who/int/csr/disease/avian_influenza/country/cases_table_2010_12_09/en/index.html.

Duffy, S., L. A. Shackelton, and E. C. Holmes. "Rates of Evolutionary Change in Viruses: Patterns and Determinants." *Nature Reviews Genetics* 9 (2008): 267–76; doi: 10.1038/nrg2323.

"Epidemiology of WHO-Confirmed Human Cases of Avian A (H5N1) Infection." *Weekly Epidemiological Record* (*WER*) 81, no. 26 (2006): 249–60.

"Historical Estimates of World Population." International Programs, U.S. Census Bureau; www.census.gov/ipc/www/worldhis.html.

Johnson, N. P., and J. Mueller. "Updating the Accounts: Global Mortality of the 1918–1920 'Spanish' Influenza Pandemic." *Bulletin of the History of Medicine* 76, no. 1 (2002): 105–15; doi:10.1353/bhm.2002.0022.

Lynn, J. "WHO Maintains 2 Billion Estimate for Likely H1N1 Cases" Reuters.com, August 4, 2009.

Newton, P., and N. D. Wolfe. "Can Animals Teach Us Medicine?" *British Medical Journal* 305 (1992): 1517–18.

Patterson, K. D., and G. F. Pyle. "The Geography and Mortality of the 1918 Influenza Pandemic." *Bulletin of the History of Medicine* 65, no. 1 (Spring 1991): 4–21.

Sarkees, M. R. "The Correlates of War Data on War: An Update to 1997." *Conflict Management and Peace Science* 18, no. 1 (2000): 123–44; doi.10.1177/073889420001800105.

Sipress, A. "Thai Boy Dies from Bird Flu; Indonesia Reports Spread of Virus." *Washington Post*, January 26, 2004, final edition.

Small, M., and J. D. Singer. *Resort to Arms: International and Civil Wars, 1816–1980.* Beverly Hills, Calif.: Sage Publications, 1982.

Taubenberger J. K., and D. M. Morens. "1918 Influenza: The Mother of All Pandemics." Centers for Disease Control and Prevention (January 2006).

"Transmission Dynamics and Impact of Pandemic Influenza A (H1N1) 2009 Virus." *Weekly Epidemiological Record* (*WER*) 84, no. 46 (2009): 481–84.

1: THE VIRAL PLANET

Acheson, N. H. *Fundamentals of Molecular Virology.* New York: Wiley, 2006: 4.

Bergh, O., K. Y. Borsheim, G. Bratbak, and M. Heldal. "High Abundance of Viruses Found in Aquatic Environments." *Nature* 340 (1989): 467–68; doi:10.1038/340467a0.

Burchell, A., R. Winer, S. De Sanjose, and E. Franco. "Chapter 6: Epidemiology and Transmission Dynamics of Genital HPV Infection." *Vaccine* 24 (2006): S52–61; doi:10.1016/j.vaccine.2006.05.031.

*Davies, P. *The Eerie Silence: Renewing Our Search for Alien Intelligence.* New York: Houghton Mifflin Harcourt, 2010.

Diamond, J., and N. Wolfe. "Where Will the Next Pandemic Emerge?" *Discover* (November 2008).

Domingo, E., et al. "Quasispecies Structure and Persistence of RNA Viruses." *Emerging Infectious Diseases* 4, no. 4 (1998): 521–27.

*Ewald, P. W. *Evolution of Infectious Disease.* New York: Oxford University Press, 1994.

Fuhrman, J. A. "Marine Viruses and Their Biogeochemical and Ecological Effects." *Nature* 399, no. 6736 (1999): 541–48; doi:10.1038/21119.

Garnham, P. C. C., et al. "A Strain of Plasmodium Vivax Characterized by Prolonged Incubation: Morphological and Biological Characteristics." *Bulletin of the World Health Organization* 52, no. 1 (1975): 21–32.

"Genital HPV Infection—Fact Sheet." Sexually Transmitted Diseases (STDs). Centers for Disease Control and Prevention (November 24, 2009); www.cdc.gov/std/hpv/stdfact-hpv.htm.

Holmes, E. C. "Error Thresholds and the Constraints to RNA Virus Evolution." *Trends in Microbiology* 11, no. 12 (2003): 543–46; doi:10.1016/jitim.2003.10.006.

Hulden, L., L. Hulden, and K. Heliovaara. "Natural Relapses in

Vivax Malaria Induced by Anopheles Mosquitoes." *Malaria Journal* 7, no. 1 (2008): 64–74; doi:10.1186/1475-2875-7-64.

Middelboe, M., and N. O. G. Jorgensen. "Viral Lysis of Bacteria: An Important Source of Dissolved Amino Acids and Cell Wall Compounds." *Journal of the Marine Biological Association of the United Kingdom* 86 (2006): 605–12; doi:10.1017/S0025315406013518.

Rohwer, F., and R. Vega Thurber. "Viruses Manipulate the Marine Environment." *Nature* 459, no. 7244 (2009): 207–12; doi:10.1038/nature08060.

Smith Hughes, S. "Beijerick, Martinus Willem." *Complete Dictionary of Scientific Biography.* New York: Charles Scribner's Sons, 2008: 13–15.

Vyas, A., S. K. Kim, and R. M. Sapolsky. "The Effects of Toxoplasma Infection on Rodent Behavior Are Dependent on Dose of the Stimulus." *Neuroscience* 148, no. 2 (2007): 342–48; doi:10.1016/j.neuroscience.2007.06.021.

Webster J. P., P. H. L. Lamberton, C. A. Donnelly, and E. F. Torrey. "Parasites as Causative Agents of Human Affective Disorders?: The Impact of Anti-Psychotic and Anti-Protozoan Medication on Toxoplasma gondii's Ability to Alter Host Behaviour." *Proceedings of the Royal Society B: Biological Sciences*, 273 (2006): 1023–30; doi:10.1098/rspb.2005.3413.

Wolfe, N. *The Aliens Among Us. What's Next? Dispatches on the Future of Science.* Ed. Max Brockman. New York: Vintage, 2009: 185–96.

2: THE HUNTING APE

Bailes, E., F. Gao, F. Bibollet-Ruche, V. Courgnaud, M. Peeters, P. A. Marx, Beatrice H. Hahn, and Paul M. Sharp. "Hybrid Origin of SIV in Chimpanzees." *Science* 300, no. 5626 (2003): 1713; doi:10.1126/science.1080657.

*Diamond, J. M. *The Third Chimpanzee: The Evolution and Future of the Human Animal.* New York: HarperCollins, 1992.

Hohmann, G., and B. Fruth. "New Records on Prey Capture and Meat Eating by Bonobos at Lui Kotale, Salonga National Park, Democratic Republic of Congo." *Folia Primatologica* 79, no. 2 (2008): 103–10; doi:10.1159/000110679.

Keele, B. F., et al. "Increased Mortality and AIDS-like Immunopathology in Wild Chimpanees Infected with SIVcpz." *Nature* 460, no. 7254 (2009): 515–19; doi:10.1038/nature08200.

Lee-Thorp, J. A., M. Sponheimer, B. H. Passey, D. J. De Ruiter, and T. E. Cerling. "Stable Isotopes in Fossil Hominin Tooth Enamel Suggest a Fundamental Dietary Shift in the Pliocene." *Philosophical Transactions of the Royal Society B: Biological Sciences* 365, no. 1556 (2010): 3389–96; doi:10.1098/rstb.2010.0059.

McGrew, W. C. "Savanna Chimpanzees Dig for Food." *Proceedings of the National Academy of Sciences* 104, no. 49 (2007): 19167–68; doi:10.1073/pnas.0710330105.

McPherron, S. P., Z. Alemseged, C. W. Marean, J. G. Wynn, D. Reed, D. Geraads, R. Bobe, and H. A. Bearat. "Evidence for Stone-Tool-Assisted Consumption of Animal Tissues before 3.39 Million Years Ago at Dikika, Ethiopia." *Nature* 466 (2010): 857–60; doi:10.1038/nature09248.

Sponheimer, M., and J. A. Lee-Thorp. "Isotopic Evidence for the Diet of an Early Hominid, Australopithecus Africanus." *Science* 283 (1999): 368–70; doi:10.1126/science.283.5400.368.

*Stanford, C. B. *The Hunting Apes: Meat Eating and the Origins of Human Behavior.* Princeton: Princeton University Press, 1999.

Wolfe, N. "Preventing the Next Pandemic." *Scientific American* (April 2009): 76–81.

Wolfe, N. D., C. Panosian Dunavan, and J. Diamond. "Origins of Major Human Infectious Diseases." *Nature* 447, no. 7142 (2007): 279–83; doi:10.1038/nature05775.

Wolfe, N. D., et al. "Deforestation, Hunting and the Ecology of Microbial Emergence." *Global Change and Human Health* 1, no. 1 (2000): 10–25; doi:10.1023/A:1011519513354.

Wrangham, R., M. Wilson, B. Hare, and N. D. Wolfe. "Chimpanzee Predation and the Ecology of Microbial Exchange." *Microbial Ecology in Health and Disease* 12, no. 3 (2000): 186–88; doi:10.1080/089106000750051855.

3: THE GREAT PATHOGEN BOTTLENECK

Behar, D. M., R. Villems, H. Soodyall, J. Blue-Smith, L. Pereira, E. Metspalu, R. Scozzari, H. Makkan, S. Tzur, and D. Comas. "The Dawn of Human Matrilineal Diversity." *American Journal of Human Genetics* 82, no. 5 (2008): 1130–40; doi:10.1016/j.ajhg.2008.04.002.

Black, F. L. "Infectious Diseases in Primitive Societies." *Science* 187 (1975): 515–18; doi:10.1126/science.163483.

*Cela-Conde, C. J., and F. J. Ayala. *Human Evolution: Trails from the Past.* Oxford: Oxford University Press, 2007.

*Coatney, G. R., W. E. Collins, M. Warren, and P. G. Contacos. *The Primate Malarias.* Bethesda, Md.: US Department of Health, Education, and Welfare, 1971.

Cornuet, J. M., and G. Luikart. "Description and Power Analysis of Two Tests for Detecting Recent Population Bottlenecks from Allele Frequency Data." *Genetics* 144 (1996): 2001–14.

Dobson, A. P., and E. R. Carper. "Infectious Diseases and Human Population History." *BioScience* 46, no. 2 (1996): 115–26.

Gao, F., et al. "Origin of HIV-1 in the Chimpanzee Pan Troglodytes Troglodytes." *Nature* 397, no. 6718 (1999): 436–41; doi:10.1038/17130.

Gibbons, A. "Pleistocene Population Explosions." *Science* 262, no. 5130 (1993): 27–28; doi:10.1126/science.262.5130.27.

*Goodall, J. *The Chimpanzees of Gombe: Patterns of Behavior.* Cambridge, Mass.: Belknap, 1986.

Huff, C. D., J. Xing, A. R. Rogers, D. Witherspoon, and L. B. Jorde. "Mobile Elements Reveal Small Population Size in the Ancient Ancestors of Homo Sapiens." *Proceedings of the National Academy of Sciences* 107, no. 5 (2010): 2147–52; doi:10.1073/pnas.0909000107.

*Kingdon, J. *The Kingdon Field Guide to African Mammals.* San Diego: Academic, 1997.

Liu, Weimin, et al. "Origin of the Human Malaria Parasite Plasmodium Falciparum in Gorillas." *Nature* 467, no. 7314 (2010): 420–25; doi:10.1038/nature09442.

Prugnolle, F., et al. "African Great Apes Are Natural Hosts of Multiple Related Malaria Species, including Plasmodium Falciparum."

Proceedings of the National Academy of Sciences 107, no. 4 (2010): 1458–63; doi:10.1073/pnas.0914440107.

Reed, K. E. "Early Hominid Evolution and Ecological Change through the African Plio-Pleistocene." *Journal of Human Evolution* 32 (1997): 289–322; doi:10.1006/jhev.1996.0106.

Rich, S. M., F. H. Leendertz, G. Xu, M. LeBreton, C. F. Djoko, M. N. Aminake, E. E. Takang, J. L. D. Diffo, B. L. Pike, B. M. Rosenthal, P. Formenty, C. Boesch, F. J. Ayala, and N. D. Wolfe. "The Origin of Malignant Malaria." *Proceedings of the National Academy of Sciences* 106, no. 35 (2009): 14902–7; doi:10.1073/pnas.0907740106.

Wolfe, N. D., C. Panosian Dunavan, and J. Diamond. "Origins of Major Human Infectious Diseases." *Nature* 447, no. 7142 (2007): 279–83; doi:10.1038/nature05775.

Wolfe, N. D., M. N. Eitel, J. Gockowski, P. K. Muchaal, C. Nolte, A. Tassy Prosser, J. Ndongo Torimiro, S. F. Weise, and D. S. Burke. "Deforestation, Hunting and the Ecology of Microbial Emergence." *Global Change and Human Health* 1, no. 1 (2000): 10–25; doi:10.1023/A:1011519513354.

*Wrangham, R. *Catching Fire: How Cooking Made Us Human.* New York: Basic, 2009.

Yang, Z. "Likelihood and Bayes Estimation of Ancestral Population Sizes in Hominoids Using Data From Multiple Loci." *Genetics* 162 (2002): 1811–23.

4: CHURN, CHURN, CHURN

Cleaveland, L. H., M. K. Laurenson, and L. H. Taylor. "Diseases of Humans and Their Domestic Mammals: Pathogen Characteristics, Host Range and the Risk of Emergence." *Philosophical Transactions of the Royal Society B: Biological Sciences* 356 (2001): 991–99; doi:10.1098/rstb.2001.0889.

Currie, C. R., J. A. Scott, R. C. Summerbell, and D. Malloch. "Letters: Fungus-Growing Ants Use Antibiotic-Producing Bacteria to Control Garden Parasites." *Nature* 398, no. 6729 (1999): 701–4; doi:10.1038/19519.

Currie, C. R., U. G. Mueller, and D. Malloch. "The Agricultural Pathology of Ant Fungus Gardens." *Proceedings of the National*

Academy of Sciences 96 (1999): 7998–8002; doi:10.1073/pnas.96.14.7998.

Delmas, O., E. C. Holmes, C. Talbi, F. Larrous, L. Dacheux, C. Bouchier, and H. Bourhy. "Genomic Diversity and Evolution of the Lyssaviruses." *PLoS ONE* E2057 3, no. 4 (2008): 1–6; doi:10.1371/journal.pone.0002057.

Diamond, J. "Evolution, Consequences, and Future of Plant and Animal Domestication." *Nature* 418, no. 6898 (2002): 700–707; doi:10.1038/nature01019.

*Diamond, J. M. *Guns, Germs, and Steel: The Fates of Human Societies.* New York: W. W. Norton, 1999.

Epstein, J. H., H. E. Field, S. Luby, J. R. C. Pulliam, and P. Daszak. "Nipah Virus: Impact, Origins, and Causes of Emergence." *Current Infectious Disease Reports* 8, no. 1 (2006): 59–65; doi:10.1007/S11908-006-0036-2.

Fagbami, A. H., T. P. Monath, and A. Fabiyi. "Dengue Virus Infections in Nigeria: A Survey for Antibodies in Monkeys and Humans." *Transactions of the Royal Society of Tropical Medicine and Hygiene* 71, no. 1 (1977): 60–65; doi:10.1016/0035-9203(77)90210-3.

Field, H. E., J. S. Mackenzie, and P. Daszak. "Henipaviruses: Emerging Paramyxoviruses Associated with Fruit Bats." *Current Topics in Microbiology and Immunology* 315 (2007): 133–59; doi:10.1007/978-3-540-70962-6_7.

Field, H., P. Young, J. M. Yob, J. Mills, L. Hall, and J. Mackenzie. "The Natural History of Hendra and Nipah Viruses." *Microbes and Infection* 3, no. 4 (2001): 307–14; doi:10.1016/S1286-4579(01)01384-3.

*Goodall, J. *The Chimpanzees of Gombe: Patterns of Behavior.* Cambridge, Mass.: Belknap, 1986.

Holmes, E. C., and S. S. Twiddy. "The Origin, Emergence and Evolutionary Genetics of Dengue Virus." *Infection, Genetics, and Evolution* 3 (2003): 19–28; doi:10.1016/S1567-1348(03)00004-2.

Hughes, W. O. H., J. Eilenberg, and J. J. Boomsma. "Trade-Offs in Group Living: Transmission and Disease Resistance in Leaf-cutting Ants." *Proceedings of the Royal Society B: Biological Sciences* 269, no. 1502 (2002): 1811–19; doi:10.1098/rspb.2002.2113.

LeBreton, M., S. Umlauf, C. F. Djoko, P. Daszak, D. S. Burke, P. Y.

Kwenkam, and N. D. Wolfe. "Rift Valley Fever in Goats, Cameroon." *Emerging Infectious Diseases* 12, no. 4 (2006): 702–3.

Li, W., et al. "Bats Are Natural Reservoirs of SARS-Like Coronaviruses." *Science* 310, no. 5748 (2005): 676–79; doi:10.1126/science.1118391.

Luby, S. P., E. S. Gurley, and M. J. Hossain. "Transmission of Human Infection with Nipah Virus." *Clinical Infectious Diseases* 49, no. 11 (2009): 1743–48; doi:10.1086/647951.

Mackenzie, J. S., D. J. Gubler, and L. R. Petersen. "Emerging Flaviviruses: The Spread and Resurgence of Japanese Encephalitis, West Nile and Dengue Viruses." *Nature Medicine* 10, no. 12s (2004): S98–109; doi:10.1038/nm1144.

Pang, J. F., et al. "MtDNA Data Indicate a Single Origin for Dogs South of Yangtze River, Less Than 16,300 Years Ago, from Numerous Wolves." *Molecular Biology and Evolution* 26, no. 12 (2009): 2849–64; doi:10.1093/molbev/msp195.

*Panter-Brick, C., R. H. Layton, and P. Rowley-Conwy. *Hunter-Gatherers: An Interdisciplinary Perspective.* Cambridge, England: Cambridge University Press, 2001.

Rudnick, A., and T. W. Lim. "Dengue Fever Studies in Malaysia." *Institute for Medical Research Malaysia Bulletin* 23 (1986): 127–47.

VonHoldt, B. M., et al. "Letters: Genome-Wide SNP and Haplotype Analyses Reveal a Rich History Underlying Dog Domestication." *Nature* 464, no. 7290 (2010): 898–903; doi:10.1038/nature08837.

Wilson, E. O. "Caste and Division of Labor in Leaf-Cutter Ants (Hymenoptera: Formicidae: Atta)." *Behavioral Ecology and Sociobiology* 7, no. 2 (1980): 143–56.

Yob, J. M., et al. "Nipah Virus Infection in Bats (Order Chiroptera) in Peninsular Malaysia." *Emerging Infectious Diseases* 7 (2001): 439–41.

5: THE FIRST PANDEMIC

Associated Press. "Tennesse Teen Dies of Rabies." *Times Daily* [Florence, Ala.], September 2, 2002.

Bermejo, M., J. D. Rodriguez-Teijeiro, G. Illera, A. Barroso, C. Vila, and P. D. Walsh. "Ebola Outbreak Killed 5000 Gorillas." *Science* 314, no. 5805 (2006): 1564; doi:10.1126/science.1133105.

Bertrand, M. "Training without Reward: Traditional Training of Pig-Tailed Macaques as Coconut Harvesters." *Science* 155, no. 3761 (1967): 484–86; doi:10.1126/science.155.3761.484.

"CDC—Outbreak of Human Monkeypox, Democratic Republic of Congo, 1996 to 1997." Centers for Disease Control and Prevention (February 9, 2011); http://www.cdc.gov/ncidod/eid/vol7no3/hutin.htm.

Cleaveland, S., D. T. Haydon, and L. Taylor. "Overviews of Pathogen Emergence: Which Pathogens Emerge, When and Why?" *Current Topics in Microbiology and Immunology* 315 (2007): 85–111; doi: 10.1007/978-3-540-70962-6_5.

Focosi, D., et al. "Torquetenovirus Viremia Kinetics after Autologous Stem Cell Transplantation Are Predictable and May Serve as a Surrogate Marker of Functional Immune Reconstitution." *Journal of Clinical Virology* 47 (2010): 189–92; doi:10.1016/j.jcv.2009.11.027.

Halpin, K., A. D. Hyatt, R. K. Plowright, J. H. Epstein, P. Daszak, H. E. Field, L. Wang, and P. W. Daniels. "Emerging Viruses: Coming in on a Wrinkled Wing and a Prayer." *Clinical Infectious Diseases* 44, no. 5 (2007): 711–17; doi:10.1086/511078.

"Human Rabies—Tennessee, 2002." *JAMA* 288, no. 24 (2002): 828–29.

Keele, B. F., et al. "Chimpanzee Reservoirs of Pandemic and Nonpandemic HIV-1." *Science* 313, no. 5786 (2006): 523–26; doi:10.1126/science.1126531.

Ladnyj, I. D., P. Ziegler, and E. Kima. "A Human Infection Caused by Monkeypox Virus in Basankusu Territory, Democratic Republic of the Congo." *Bulletin of the World Health Organization* 46 (1972): 593–97.

Leroy, E. M., B. Kumulungui, X. Pourrut, P. Rouquet, A. Hassanin, P. Yaba, A. Délicat, J. T. Paweska, J. P. Gonzalez, and R. Swanepoel. "Fruit Bats as Reservoirs of Ebola Virus." *Nature* 438, no. 7068 (2005): 575–76; doi:10.1038/438575a.

Prescott, L. E., et al. "Correspondence: Global Distribution of Transfusion-Transmitted Virus." *New England Journal of Medicine* 339, no. 11 (1998): 776–77.

Rimoin, A. W., et al. "Major Increase in Human Monkeypox Incidence 30 Years after Smallpox Vaccination Campaigns Cease in the Democratic Republic of Congo." *Proceedings of the National Academy of Sciences* 107, no. 37 (2010): 16262–67; doi:10.1073/pnas.1005769107.

Rimoin A. W., N. Kisalu, B. Kebela-Ilunga, T. Mukaba, L. L. Wright, P. Formenty, N. D. Wolfe, R. L. Shongo, F. Tshioko, E. Okitolonda, J. J. Muyembe, R. W. Ryder, and H. Meyer. "Endemic Human Monkeypox, Democratic Republic of Congo, 2001–2004." *Emerging Infectious Disease* 13 (2007): 934–37; doi:10.3201/eid1306.061540.

Simmonds, P., F. Davidson, C. Lycett, L. Prescott, D. Macdonald, J. Ellender, P. Yap, C. Ludlam, G. Haydon, J. Gillon, and L. M. Jarvis. "Detection of a Novel DNA Virus (TT Virus) in Blood Donors and Blood Products." *Lancet* 352, no. 9123 (1998): 191–95; doi:10.1016/S0140-6736(98)03056-6.

"Update: Multistate Outbreak of Monkeypox—Illinois, Indiana, Kansas, Missouri, Ohio, and Wisconsin, 2003." Centers for Disease Control and Prevention (February 9, 2011); http://www.cdc.gov/mmwr/preview/mmwrhtml/mm5227a5.htm.

Van Blerkom, L. M. "Role of Viruses in Human Evolution." *American Journal of Physical Anthropology* 46 (2003): 14–46; doi:10.1002/ajpa.10384.

Voelker, R. "Suspected Monkeypox Outbreak." *JAMA* 279, no. 2 (1998): 101.

Wolfe, N. D., C. Panosian Dunavan, and J. Diamond. "Origins of Major Human Infectious Diseases." *Nature* 447, no. 7142 (2007): 279–83; doi:10.1038/nature05775.

Woolhouse, M. E. J., D. T. Haydon, and R. Antia. "Emerging Pathogens: The Epidemiology and Evolution of Species Jumps." *TRENDS in Ecology and Evolution* 20, no. 5 (2005): 238–43; doi:10.1016/j.tree.2005.02.009.

Zhang, W., K. Chaloner, H. Tillmann, C. Williams, and J. Stapleton. "Effect of Early and Late GB Virus C Viraemia on Survival of HIV-Infected Individuals: A Meta-analysis." *HIV Medicine* 7, no. 3 (2006): 173–80; doi:10.1111/J.1468 1293 2006.00366.X.

6: ONE WORLD

Arcadi, A. C., and R. W. Wrangham. "Infanticide in Chimpanzees: Review of Cases and a New Within-group Observation from the Kanyawara Study Group in Kibale National Park." *Primates* 40, no. 2 (1999): 337–51; doi:10.1007/BF02557557.

Arroyo, M. A., W. B. Sateren, D. Serwadda, R. H. Gray, M. J. Wawer, N. K. Sewankambo, N. Kiwanuka, G. Kigozi, F. Wabwire-Mangen, M. Eller, L. A. Eller, D. L. Birx, M. L. Robb, and F. E. McCutchan. "Higher HIV-1 Incidence and Genetic Complexity Along Main Roads in Rakai District, Uganda." *Journal of Acquired Immune Deficiency Syndromes (JAIDS)* 43, no. 4 (2006): 440–45; doi:10.1097/01.gov.0000243053.80945.f0.

Berger, L., et al. "Chytridiomycosis Causes Amphibian Mortality Associated with Population Declines in the Rain Forests of Australia and Central America." *Proceedings of the National Academy of Sciences* 95 (1998): 9031–36.

Brownstein, J. S., C. J. Wolfe, and K. D. Mandl. "Empirical Evidence for the Effect of Airline Travel on Inter-Regional Influenza Spread in the United States." *PLoS Medicine* 3, no. 10 (2006): e401; doi:10.1371/journal.pined.0030401.

Chitnis, A., D. Rawls, and J. Moore. "Origin of HIV Type 1 in Colonial French Equatorial Africa?" *AIDS Research and Human Retroviruses* 16, no. 1 (2000): 5–8; doi:10.1.089/088922200309548.

*Cook, N. D. *Born to Die: Disease and New World Conquest, 1492–1650.* Cambridge, England: Cambridge University Press, 1998.

Daszak, P., L. Berger, A. A. Cunningham, A. D. Hyatt, D. E. Green, and R. Spear. "Emerging Infectious Diseases and Amphibian Population Declines." *Emerging Infectious Diseases* 5, no. 6 (1999): 735–48.

Ducrot, C., M. Arnold, A. De Koeijer, D. Heim, and D. Calavas. "Review on the Epidemiology and Dynamics of BSE Epidemics." *Veterinary Research* 39, no. 15 (2008); doi:10.1051/vetres:2007053.

Folsom, J. "AIDS Origins: Colonial Legacies and the Belgian Congo." Paper presented at the annual meeting of the American Sociological Association Annual Meeting, Hilton Atlanta and Atlanta Marriott Marquis, Atlanta, Ga., August 13, 2010.

Graves, M. W., and D. J. Addison. "The Polynesian Settlement of the Hawaiian Archipelago: Integrating Models and Methods in Archaeological Interpretation." *World Archaeology* 26, no. 3 (1995): 380–99; doi:10.1080/00438243.1995.9980283.

Hudjashov, G., T. Kivisild, P. A. Underhill, P. Endicott, J. J. Sanchez, A. A. Lin, P. Shen, P. Oefner, C. Renfrew, R. Villems, and P. Forster. "Revealing the Prehistoric Settlement of Australia by Y Chromosome and MtDNA Analysis." *Proceedings of the National Academy of Sciences* 104, no. 21 (2007): 8726–30; doi:10.1073/pnas.0702928104.

Lawler, A. "Report of Oldest Boat Hints at Early Trade Routes." *Science* 296 (2002): 1791–92; doi:10.1126/science.296.5574.1791.

*Lay, M. G. *Ways of the World: A History of the World's Roads and of the Vehicles That Used Them.* Piscataway, N.J.: Rutgers University Press, 1992.

Morens, D. M., J. K. Taubenberger, G. K. Folkers, and A. S. Fauci. "Pandemic Influenza's 500th Anniversary." *Clinical Infectious Diseases* 51 (2010): 1442–44; doi:10.1086/657429.

Perrin, L., L. Kaiser, and S. Yerly. "Travel and the Spread of HIV-1 Genetic Variants." *Lancet Infectious Diseases* 3, no. 1 (2003): 22–27; doi:10.1016/S1473-3099(03)00484-5.

*Quammen, D. *The Song of the Dodo: Island Biogeography in an Age of Extinctions.* New York: Simon and Schuster, 1997.

Russell, C. A., et al. "The Global Circulation of Seasonal Influenza A (H3N2) Viruses." *Science* 320, no. 5874 (2008): 340–46; doi:10.1126/science.1154137.

———. "Influenza Vaccine Strain Selection and Recent Studies on the Global Migration of Seasonal Influenza Viruses." *Vaccine* 26 (2008): D31–34; doi:10.1016/j.vaccine.2008.07.078.

Tatem, A. J., D. J. Rogers, and S. I. Hay. "Global Transport Networks and Infectious Disease Spread." *Advances in Parasitology* 62 (2006): 293–343; doi:10.1016/S0065-308x(05)62009-x.

Wilson, M. "Travel and the Emergence of Infectious Diseases." *Emerging Infectious Diseases* 1, no. 2 (1995): 39–46.

Worobey, M., M. Gemmel, D. E. Teuwen, T. Haselkorn, K. Kunstman, M. Bunce, J. J. Muyembe, J. M. M. Kabongo, R. M. Kalengayi, E. Van Marck, M. T. P. Gilbert, and S. M. Wolinsky.

"Direct Evidence of Extensive Diversity of HIV-1 in Kinshasa by 1960." *Nature* 455, no. 7213 (2008): 661–64; doi:10.1038/nature 07390.

*Wrangham, R. W., W. C. McGrew, F. B. M. De Waal, and P. G. Heltne. *Chimpanzee Cultures.* Chicago, Ill.: Chicago Academy of Sciences, 1994.

Zhu, T., B. T. Korber, A. J. Nahmias, E. Hooper, P. M. Sharp, and D. D. Ho. "An African HIV-1 Sequence from 1959 and Implications for the Origin of the Epidemic." *Nature* 391, no. 6667 (1998): 594–97; doi:10.1038/35400.

7: THE INTIMATE SPECIES

Allain, J. P., S. L. Stramer, A. B. F. Carneiro-Proietti, M. L. Martins, S. N. Lopes Da Silva, M. Ribeiro, F. A. Proietti, and H. W. Reesink. "Transfusion-Transmitted Infectious Diseases." *Biologicals* 37, no. 2 (2009): 71–77; doi:10.1016/j.biologicals.2009.01 .002.

Apetrei, C., et al. "Potential for HIV Transmission through Unsafe Injections." *AIDS* 20, no. 7 (2006): 1074–76; doi:10.1097/01.aids .0000222085.21540.8a.

Atkin, S. J. L., B. E. Griffin, and S. M. Dilworth. "Polyoma Virus and Simian Virus 40 as Cancer Models: History and Perspectives." *Seminars in Cancer Biology* 19, no. 4 (2009): 211–17; doi:10 .1016/j.semcancer.2009.03.001.

Berry, N., C. Davis, A. Jenkins, D. Wood, P. Minor, G. Schild, M. Bottiger, H. Holmes, and N. Almond. "Vaccine Safety: Analysis of Oral Polio Vaccine CHAT Stocks." *Nature* 410, no. 6832 (2001): 1046–47; doi:10.1038/35074176.

Blancou, P., J. P. Vartanian, C. Christopherson, N. Chenciner, C. Basilico, S. Kwok, and S. Wain-Hobson. "Polio Vaccine Samples Not Linked to AIDS." *Nature* 410, no. 6832 (2001): 1045–46; doi:10.1038/35074171.

*Brock, P. *Charlatan: America's Most Dangerous Huckster, The Man Who Pursued Him, and the Age of Flimflam.* New York: Three Rivers, 2008.

Curtis, T. "The Origin of AIDS: A Startling New Theory Attempts to Answer the Question, 'Was It an Act of God or an Act of Man?' " *Rolling Stone*, March 19, 1992, 54–59, 61, 106, 108.

Dang-Tan, T., S. M. Mahmud, R. Puntoni, and E. L. Franco. "Polio Vaccines, Simian Virus 40, and Human Cancer: The Epidemiologic Evidence for a Causal Association." *Oncogene* 23 (2004): 6535–40; doi:10.1038/sj.onc.1207877.

Deschamps, J. Y., F. A. Roux, P. Sai, and E. Gouin. "History of Xenotransplantation." *Xenotransplantation* 12, no. 2 (2005): 91–109; doi:10.1111/j.1399-3089.2004.00199.x.

Dorfer, L., M. Moser, F. Bahr, K. Spindler, E. Egarter-Vigl, S. Giullén, G. Dohr, and T. Kenner. "A Medical Report from the Stone Age?" *Lancet* 354 (September 1999): 1023–25; doi:10.1016/S0140-6736(98)12242-0.

Drazan, K. E. "Molecular Biology of Hepatitis C Infection." *Liver Transplantation* 6 (2000): 396–406; doi:10.1053/jlts.2000.6449.

Drucker, E., P. G. Alcabes, and P. A. Marx. "The Injection Century: Massive Unsterile Injections and the Emergence of Human Pathogens." *Lancet* 358, no. 9297 (2001): 1989–92; doi:10.1016/S0140-6736(01)06967-7.

Fischer, L., M. Sterneck, M. Claus, A. Costard-Jackle, B. Fleischer, H. Herbst, X. Rogiers, and C. E. Broelsch. "Transmission of Malaria Tertiana by Multi-Organ Donation." *Clinical Transplantation* 13 (1999): 491–95; doi:10.1034/j.1399-0012.1999.130609.x.

Herwaldt, B. L. "Laboratory-Acquired Parasitic Infections from Accidental Exposures." *Clinical Microbiology Reviews* 14, no. 4 (2001): 659–88; doi:10.1128/CMR.14.3.659-688.2001.

Houff, S. A., et al. "Human-to-Human Transmission of Rabies Virus by Corneal Transplant." *New England Journal of Medicine* 300 (1979): 603–4.

Kahn, A. "Regaining Lost Youth: The Controversial and Colorful Beginnings of Hormone Replacement Therapy in Aging." *Journal of Gerontology: Biological Sciences* 60A, no. 2 (2005): 142–47; doi:10.1093/gerona/60.2.142.

Korber, B., M. Muldoon, J. Theiler, F. Gao, R. Gupta, A. Lapedes, B. H. Hahn, S. Wolinsky, and T. Bhattacharya. "Timing the Ancestor of the HIV-1 Pandemic Strains." *Science* 288, no. 5472 (2000): 1789–96; doi:10.1126/science.288.5472.1789.

Marx, P. A., P. G. Alcabes, and E. Drucker. "Serial Human Passage of Simian Immunodeficiency Virus by Unsterile Injections and the Emergence of Epidemic Human Immunodeficiency Virus in Africa." *Philosophical Transactions of the Royal Society B: Biological Sciences* 356, no. 1410 (2001): 911–20; doi:10.1098/rstb.2001.0867.

Mejia, G. A., et al. "Malaria in a Liver Transplant Recipient: A Case Report." *Transplantation Proceedings* 38, no. 9 (2006): 3132–34; doi:10.1016/j.transproceed.2006.08.187.

Morgenthaler, J. J. "Securing Viral Safety for Plasma Derivatives." *Transfusion Medicine Reviews* 15, no. 3 (2001): 224–33; doi:10.1053/tmrv.2001.24590.

Palacios, G., et al. "A New Arenavirus in a Cluster of Fatal Transplant-Associated Diseases." *New England Journal of Medicine* 358, no. 10 (2008): 991–98.

Paradis, K., G. Langford, Z. Long, W. Heneine, P. Sandstrom, W. M. Switzer, L. E. Chapman, C. Lockey, D. Onions, and E. Otto. "Search for Cross-Species Transmission of Porcine Endogenous Retrovirus in Patients Treated with Living Pig Tissue." *Science* 285, no. 5431 (1999): 1236–41; doi:10.1126/science.285.5431.1236.

Pike, R. M. "Laboratory-Associated Infections: Incidence, Fatalities, Causes, and Prevention." *Annual Review of Microbiology* 33, no. 1 (1979): 41–66; doi:10.1146/annurev.mi.33.100179.000353.

"Pneumocystis Pneumonia—Los Angeles." *MMWR* 30, no. 21 (June 5, 1981): 250–52.

Poinar, H., M. Kuch, and S. Pääbo. "Molecular Analyses of Oral Polio Vaccine Samples." *Science* 292, no. 5517 (2001): 743–44; doi:10.1126/science.1058463.

Pybus, O. G., P. V. Markov, A. Wu, and A. J. Tatem. "Investigating the Endemic Transmission of the Hepatitis C Virus." *International Journal for Parasitology* 37 (2007): 839–49; doi:10.1016/j.ijpara.2007.04.009.

Rambaut, A. L. "Human Immunodeficiency Virus: Phylogeny and the Origin of HIV-1." *Nature* 410 (2001): 1047–48; doi:10.1038/35074179.

Rudolf, V. H. W., and J. Antonovics. "Disease Transmission by Cannibalism: Rare Event or Common Occurrence?" *Proceedings of the Royal Society B: Biological Sciences* 274, no. 1614 (2007): 1205–10; doi:10.1098/rspb.2006.0449.

Schreiber, G. B., M. P. Busch, S. H. Kleinman, and J. J. Korelitz. "The Risk of Transfusion-Transmitted Viral Infections." *New England Journal of Medicine* 334 (1996): 1685–90.

Sewell, D. L. "Laboratory-Associated Infections and Biosafety." *Clinical Microbiology Reviews* 8, no. 3 (1995): 389–405.

Simmonds, P. "Reconstructing the Origins of Human Hepatitis Viruses." *Philosophical Transactions of the Royal Society B: Biological Sciences* 356, no. 1411 (2001): 1013–26; doi:10.1098/rstb.2001.0890.

Smith, D. B., S. Pathirana, F. Davidson, E. Lawlor, J. Power, P. L. Yap, and P. Simmonds. "The Origin of Hepatitis C Virus Genotypes." *Journal of General Virology* 78 (1997): 321–28.

*Specter, M. *Denialism: How Irrational Thinking Hinders Scientific Progress, Harms the Planet, and Threatens Our Lives.* New York: Penguin, 2009.

Srinivasan, A., et al. "Transmission of Rabies Virus from an Organ Donor to Four Transplant Recipients." *New England Journal of Medicine* 352, no. 11 (2005): 1103–11.

"The Tattoos." Ötzi the Iceman. South Tyrol Museum of Archaeology, 2008; http://www.iceman.it/en/node/262.

Victoria, J. G., C. Wang, M. S. Jones, C. Jaing, K. McLoughlin, S. Gardner, and E. L. Delwart. "Viral Nucleic Acids in Live-Attenuated Vaccines: Detection of Minority Variants and an Adventitious Virus." *Journal of Virology* 84, no. 12 (2010): 6033–40; doi:10.1128/JVI.02690-09.

"Voronoff Patient Tells of New Life." *New York Times*, October 6, 1922.

8: VIRAL RUSH

Alpers, M. P. "Review. The Epidemiology of Kuru: Monitoring the Epidemic from Its Peak to Its End." *Philosophical Transactions of the Royal Society B: Biological Sciences* 363, no. 1510 (2008): 3707–13; doi:10.1098/rstb.2008.0071.

Ambrus, J. L., Sr., and J. L. Ambrus, Jr. "Nutrition and Infectious Diseases in Developing Countries and Problems of Acquired Immunodeficiency Syndrome." *Experimental Biology and Medicine* 229 (2004): 464–72.

Associated Press. "Century-Old Smallpox Scabs in N. M. Envelope." USATODAY.com, December 26, 2003.

Bellamy, R. J., and A. R. Freedman. "Bioterrorism." *Quarterly Journal of Medicine* 94 (2001): 227–34; doi:10.1093/9jmed/94.4.227.

Burke, D. S. "Recombination in HIV: An Important Viral Evolutionary Strategy." *Emerging Infectious Diseases* 3, no. 3 (1997): 253–59.

De Rosa, P. *Vicars of Christ: The Dark Side of the Papacy.* New York: Crown, 1988.

Ducrot, C., M. Arnold, A. De Koeijer, D. Heim, and D. Calavas. "Review on the Epidemiology and Dynamics of BSE Epidemics." *Veterinary Research* 39, no. 4 (2008): 15; doi:10.1051/vetres:2007053.

Fishbein, L. "Transmissible Spongiform Encephalopathies, Hypotheses and Food Safety: An Overview." *Science of the Total Environment* 217, nos. 1-2 (1998): 71–82; doi:10.1016/50048-9697(98)00164-8.

*Garrett, L. *The Coming Plague: Newly Emerging Diseases in a World Out of Balance.* New York: Penguin, 1995.

Harman, J. L., and C. J. Silva. "Bovine Spongiform Encephalopathy." *Journal of the American Veterinary Medical Association* 234, no. 1 (2009): 59–72.

LeBreton, M., O. Yang, U. Tamoufe, E. Mpoudi-Ngole, J. N. Torimiro, C. F. Djoko, J. K. Carr, A. Tassy Prosser, A. W. Rimoin, D. L. Birx, D. S. Burke, and N. D. Wolfe. "Exposure to Wild Primates among HIV-infected Persons." *Emerging Infectious Diseases* 13, no. 10 (2007): 1579–82.

Li, Y., et al. "Predicting Super Spreading Events during the 2003 Severe Acute Respiratory Syndrome Epidemics in Hong Kong and Singapore." *American Journal of Epidemiology* 160, no. 8 (2004): 719–28; doi:10.1093/aje/kwh273.

Morris, J. G., Jr., and M. Potter. "Emergence of New Pathogens as a

Function of Changes in Host Susceptibility." *Emerging Infectious Diseases* 3, no. 4 (1997): 435–41.

Ratnasingham, S., and P. D. N. Hebert. "Bold: The Barcode of Life Data System (http://www.barcodinglife.org)." *Molecular Ecology Notes* 7, no. 3 (2007): 355–64; doi:10.1111/j.1471-8286.2007.01678.x.

Rees, M. J. *Our Final Hour: A Scientist's Warning: How Terror, Error, and Environmental Disaster Threaten Humankind's Future in This Century on Earth and Beyond.* New York: Basic, 2003.

Reshetin, V. P., and J. L. Regens. "Simulation Modeling of Anthrax Spore Dispersion in a Bioterrorism Incident." *Risk Analysis* 23, no. 6 (2003): 1135–45; doi:10.1111/j.0272-4332.2003.00387.x.

Rusnak, J. M., M. G. Kortepeter, R. J. Hawley, A. O. Anderson, E. Boudreau, and E. Eitzen. "Risk of Occupationally Acquired Illness from Biological Threat Agents in Unvaccinated Laboratory Workers." *Biosecurity and Bioterrorism: Biodefense Strategy, Practice, and Science* 2, no. 4 (2004): 281–93; doi:10.1089/bsp.2004.2.281.

Schwartz, J. "Fish Tale Has DNA Hook: Students Find Bad Labels." *New York Times*, August 21, 2008, Science sec.

Sidel, V. W., H. W. Cohen, and R. M. Gould. "Good Intentions and the Road to Bioterrorism Preparedness." *American Journal of Public Health* 91, no. 5 (2001): 716–18.

Silbergeld, E. K., J. Graham, and L. B. Price. "Industrial Food Animal Production, Antimicrobial Resistance, and Human Health." *Annual Review of Public Health* 29, no. 1 (2008): 151–69; doi:10.1146/annurev.pubhealth.29.620907.090904.

Snyder, J. W. "Role of the Hospital-Based Microbiology Laboratory in Preparation for and Response to a Bioterrorism Event." *Journal of Clinical Microbiology* 41, no. 1 (2003): 1–4; doi:10.1128/JCM.41.1.1-4.2003.

Trevejo, R. T., M. C. Barr, and R. A. Robinson. "Important Emerging Bacterial Zoonotic Infections Affecting the Immunocompromised." *Veterinary Research* 36, no. 3 (2005): 493–506; doi:10.1051/vetres:2005011.

Tucker, J. B. "Historical Trends Related to Bioterrorism: An Empirical Analysis." *Emerging Infectious Diseases* 5, no. 4 (1999): 498–504.

UN-Habitat, United Nations Human Settlements Programme. *State of the World's Cities, 2010/2011—Cities for All: Bridging the Urban Divide*, 2010.

"Update: Multistate Outbreak of Monkeypox—Illinois, Indiana, Kansas, Missouri, Ohio, and Wisconsin, 2003." Centers for Disease Control and Prevention, February 9, 2011; http://www.cdc.gov/mmwr/preview/mmwrhtml/mm5227a5.htm.

Wang, L. F., and B. T. Eaton. "Bats, Civets, and the Emergence of SARS." *Current Topics in Microbiology and Immunology* 315 (2007): 325–44; doi:10.1007/978-3-540-70962-6_13.

Wolfe, N. D., P. Daszak, M. Kilpatrick, and D. S. Burke. "Bushmeat Hunting, Deforestation, and Prediction of Zoonotic Disease Emergence." *Emerging Infectious Diseases* 11, no. 12 (2005): 1822–27.

Wolfe, N. D., M. Ngole Eitel, J. Gockowski, P. K. Muchaal, C. Nolte, A. Tassy Prosser, J. Ndongo Torimiro, S. F. Weise, and D. S. Burke. "Deforestation, Hunting and the Ecology of Microbial Emergence." *Global Change and Human Health* 1, no. 1 (2000): 10–25; doi:10.1023/A:1011519513354.

Woolhouse, M. E. J., R. Howey, E. Gaunt, L. Reilly, M. Chase-Topping, and N. Savill. "Temporal Trends in the Discovery of Human Viruses." *Proceedings of the Royal Society B: Biological Sciences* 275, no. 1647 (2008): 2111–15; doi:10.1098/rspb.2008.0294.

Zhang, X. W., Y. L. Yap, and A. Danchin. "Testing the Hypothesis of a Recombinant Origin of the SARS-Associated Coronavirus." *Archives of Virology* 150, no. 1 (2005): 1–20; doi:10.1007/s00705-004-0413-9.

Zimmer, S. M., and D. S. Burke. "Historical Perspective—Emergence of Influenza A (H1N1) Viruses." *New England Journal of Medicine* 361, no. 3 (2009): 279–85; doi:10.1056/NEJMra0904322.

9: VIRUS HUNTERS

Botten, J., K. Mirowsky, C. Ye, K. Gottlieb, M. Saavedra, L. Ponce, and B. Hjelle. "Shedding and Intracage Transmission of Sin Nombre Hantavirus in the Deer Mouse (*Peromyscus maniculatus*)

Model." *Journal of Virology* 76, no. 15 (2002): 7587–94; doi:10.1128/JVI.76.15.7587-7594.2002.

Carr, J. K., N. D. Wolfe, J. N. Torimiro, U. Tamoufe, E. Mpoudi-Ngole, L. Eyzaguirre, D. L. Birx, F. E. McCutchan, and D. S. Burke. "HIV-1 Recombinants with Multiple Parental Strains in Low-Prevalence, Remote Regions of Cameroon: Evolutionary Relics?" *Retrovirology* 7 (2010): 39; doi:10.1186/1742-4690-7-39.

Deblauwe, I., P. Guislain, J. Dupain, and L. Van Elsacker. "Use of a Tool-Set by *Pan troglodytes troglodytes* to Obtain Termites (*Macrotermes*) in the Periphery of the Dja Biosphere Reserve, Southeast Cameroon." *American Journal of Primatology* 68, no. 12 (2006): 1191–96; doi:10.1002/ajp.20318.

Formenty, P., C. Hatz, B. Le Guenno, A. Stoll, P. Rogenmoser, and A. Widmer. "Human Infection Due to Ebola Virus, Subtype Côte D'Ivoire: Clinical and Biologic Presentation." *Journal of Infectious Diseases* 179, no. S1 (1999): S48–53; doi:10.1086/514285.

Kalish M. L., N. D. Wolfe, C. Ndongmo, C. Djoko, K. E. Robbins, J. McNicholl, M. Aidoo, P. Fonjungo, G. Alemnji, E. Mpoudi-Ngole, C. Zeh, T. M. Folks, and D. S. Burke. "Central African Hunters Exposed to SIV." *Emerging Infectious Diseases* 11 (2005): 1928–30.

Keefe, P. R. *Chatter: Dispatches from the Secret World of Global Eavesdropping.* New York: Random House, 2005.

*Lee, R. B., and I. DeVore. *Man the Hunter.* Piscataway, N.J.: Transaction, 1968.

Leendertz, F. H., et al. "Anthrax in Western and Central African Great Apes." *American Journal of Primatology* 68, no. 9 (2006): 928–33; doi:10.1002/ajp.20298.

——. "Anthrax Kills Wild Chimpanzees in a Tropical Rainforest." *Nature* 430 (2004): 451–52; doi:10.1038/nature02722.

Pike, B. L., K. E. Saylors, J. N. Fair, M. LeBreton, U. Tamoufe, C. F. Djoko, A. W. Rimoin, and N. D. Wolfe. "The Origin and Prevention of Pandemics." *Clinical Infectious Diseases* 50, no. 12 (2010): 1636–40; doi:10.1086/652860.

Pruetz, J. D., and P. Bertolani. "Savanna Chimpanzees, *Pan troglodytes verus*, Hunt with Tools." *Current Biology* 17, no. 5 (2007): 412–17; doi:10.1016/j.cub.2006.12.042.

Switzer, W. M., I. Hewlett, L. Aaron, N. D. Wolfe, D. S. Burke, T. M. Folks, and W. Heneine. "Serological Testing for Human T-lymphotropic Virus-3 and -4." *Transfusion* 46 (2006): 1647–48; doi:10.1111/j.1537-2995.2006.00950.X.

Switzer, W. M., M. Salemi, V. Shanmugam, F. Gao, M. Cong, C. Kuiken, V. Bhullar, B. E. Beer, D. Vallet, A. Gautier-Hion, Z. Tooze, F. Villinger, E. C. Holmes, and W. Heneine. "Ancient Co-Speciation of Simian Foamy Viruses and Primates." *Nature* 434, no. 7031 (2005): 376–80; doi:10.1038/nature03341.

National Commission on Terrorist Attacks Upon the United States. *The 9/11 Commission Report*. Washington, D.C.: National Commission on Terrorist Attacks Upon the United States, 2004.

Wolfe, N. D., et al. "Emergence of Unique Primate T-lymphotropic Viruses among Central African Bushmeat Hunters." *Proceedings of the National Academy of Sciences* 102, no. 22 (2005): 7994–99; doi:10.1073/pnas.0501734102.

——. "Naturally Acquired Simian Retrovirus Infections in Central African Hunters." *Lancet* 363, no. 9413 (2004): 932–37; doi:10.1016/S0140-6736(04)15787-5.

Wolfe, N. D., W. M. Switzer, T. M. Folks, D. S. Burke, and W. Heneine. "Simian Retroviral Infections in Human Beings." *Lancet* 364 (2004): 139–40 [letter].

Wolfe, N. D., A. A. Escalante, W. B. Karesh, A. Kilbourn, A. Spielman, and A. A. Lal. "Wild Primate Populations in Emerging Infectious Disease Research: The Missing Link?" *Emerging Infectious Diseases* 4, no. 2 (1998): 149–58.

Wolfe, N. D., A. T. Prosser, J. K. Carr, U. Tamoufe, E. Mpoudi-Ngole, J. N. Torimiro, M. LeBreton, F. E. McCutchan, D. L. Birx, and D. S. Burke. "Exposure to Nonhuman Primates in Rural Cameroon." *Emerging Infectious Diseases* 10 (2004): 2094–99.

10: MICROBE FORECASTING

Brownstein, J. S., C. C. Freifeld, and L. C. Madoff. "Perspective: Digital Disease Detection—Harnessing the Web for Public Health Surveillance." *New England Journal of Medicine* 360, no. 21 (2009): 2153–57; doi:10.1056/NEJMP0900702.

Christakis, N. A., and J. H. Fowler. "Social Network Sensors for Early Detection of Contagious Outbreaks." *PLoS ONE* e12948 5, no. 9 (2010); doi:10.1371/journal.pone.0012948.

Dezsö, Z., and A.-L. Barabási. "Halting Viruses in Scale-Free Networks." *Physical Review E* 65, no. 5 (2002); doi:10.1103/PhysRevE.65.055103.

Eagle, N., A. Pentland, and D. Lazer. "Inferring Friendship Network Structure by Using Mobile Phone Data." *Proceedings of the National Academy of Sciences* 106, no. 36 (2009): 15274–78; doi:10.1073/pnas.0900282106.

Freifeld, C. C., Rumi C., S. R. Mekaru, E. H. Chan, T. Kass-Hout, A. Ayala Iacucci, and J. S. Brownstein. "Participatory Epidemiology: Use of Mobile Phones for Community-Based Health Reporting." *PLoS Medicine* e1000376 7, no. 12 (2010); doi:10.1371/journal.pined.1000376.

Ginsberg, J., M. H. Mohebbi, R. S. Patel, L. Brammer, M. S. Smolinski, and L. Brilliant. "Detecting Influenza Epidemics Using Search Engine Query Data." *Nature* 457, no. 7232 (2008): 1012–14; doi:10.1038/nature07634.

Gómez-Sjöberg, R., A. A. Leyrat, D. M. Pirone, C. S. Chen, and S. R. Quake. "Versatile, Fully Automated, Microfluidic Cell Culture System." *Analytical Chemistry* 79, no. 22 (2007): 8557–63; doi:10.1021/ac071311w.

González, M. C., C. A. Hidalgo, and A.-L. Barabási. "Understanding Individual Human Mobility Patterns." *Nature* 453, no. 7196 (2008): 779–82; doi:10.1038/nature06958.

Kapoor, A., N. Eagle, and E. Horvitz. "People, Quakes, and Communications: Inferences from Call Dynamics about a Seismic Event and Its Influences on a Population." *Association for the Advancement of Artificial Intelligence Spring Symposium Series* (2010): 51–56.

Lampos, V., and N. Cristianini. "Tracking the Flu Pandemic by Monitoring the Social Web." *2nd International Workshop on Cognitivie Information Processing* (2010): 411–16; doi:10.1109/CIP.2010.5604088.

Lauring, A. S., J. O. Jones, and R. Andino. "Rationalizing the Development of Live Attenuated Virus Vaccine." *Nature Biotechnology* 28, no. 6 (2010): 573–79; doi:10.1038/nbt.1635.

Lauring, A. S., and R. Andino. "Quasispecies Theory and the Behavior of RNA Viruses." *PLoS Pathogens* e1001005 6, no. 7 (2010); doi:10.1371/journal.ppat.1001005.

Lipkin, W. I. "Microbe Hunting." *Microbiology and Molecular Biology Reviews* 74, no. 3 (2010): 363–77.

Meyers, L. A., M. E. J. Newman, and B. Pourbohloul. "Predicting Epidemics on Directed Contact Networks." *Journal of Theoretical Biology* 240, no. 3 (2006): 400–18; doi:10.1016/j.tbi.2005.10.004.

Palacios, G., J. Druce, L. Du, T. Tran, C. Birch, T. Briese, S. Conlan, P. L. Quan, J. Hui, J. Marshall, J. F. Simons, M. Egholm, C. D. Paddock, W. J. Shieh, C. S. Goldsmith, S. R. Zaki, M. Catton, and W. I. Lipkin. "A New Arenavirus in a Cluster of Fatal Transplant-Associated Diseases." *New England Journal of Medicine* 358, no. 10 (2008): 991–98; doi:10.1056/NEJMoa073785.

Paneth, N. "Assessing the Contributions of John Snow to Epidemiology: 150 Years After Removal of the Broad Street Pump Handle." *Epidemiology* 15, no. 5 (2004): 514–16; doi:10.1097/01.ede.0000135915.94799.00.

Pastor-Satorras, R., and A. Vespignani. "Epidemic Dynamics and Endemic States in Complex Networks." *Physical Review E* 63, no. 6 (2001); doi:10.1103/PhysRevE.63.066117.

Polgreen, P. M., F. D. Nelson, and G. R. Neumann. "Healthcare Epidemiology: Use of Prediction Markets to Forecast Infectious Disease Activity." *Clinical Infectious Diseases* 44, no. 2 (2007): 272–79; doi:10.1086/510427.

Quan, P.-L., T. Briese, G. Palacios, and W. I. Lipkin. "Rapid Sequence-Based Diagnosis of Viral Infection." *Antiviral Research* 79, no. 1 (2008): 1–5; doi:10.1016/j.antiviral.2008.02.002.

Safaie, A., S. M. Mousavi, R. E. LaPorte, M. M. Goya, and M. Zahraie. "Introducing a Model for Communicable Diseases Surveillance: Cell Phone Surveillance (CPS)." *European Journal of Epidemiology* 21, no. 8 (2006): 627–32; doi:10.007/s10654-006-9033-x.

Shalon, D., S. J. Smith, and P. O. Brown. "A DNA Microarray System for Analyzing Complex DNA Samples Using Two-color Fluorescent Probe Hybridization." *Genome Research* 6, no. 7 (1996): 639–45; doi:10.1101/gr.6.7.639.

Snow, J. "Cholera, and the Water Supply in the South Districts of London." *British Medical Journal* 1, no. 42 (1857): 864–65.

Vance, K., W. Howe, and R. P. Dellavalle. "Social Internet Sites as a Source of Public Health Information." *Dermatologic Clinics* 27, no. 2 (2009): 133–36; doi:10.1016/j.det.2008.11.010.

Tang, P., and C. Chiu. "Metagenomics for the Discovery of Novel Human Viruses." *Future Microbiology* 5, no. 2 (2010): 177–89; doi:10.2217/fmb.09.120.

Wang, D., A. Urisman, Y.-T. Liu, M. Springer, T. G. Ksiazek, D. D. Erdman, E. R. Mardis, M. Hickenbotham, V. Magrini, J. Eldred, J. P. Latreille, R. K. Wilson, D. Ganem, and J. L. DeRisi. "Viral Discovery and Sequence Recovery Using DNA Microarrays." *PLoS Biology* 1, no. 2 (2003): e2; doi:10.1371/journal.pbio.0000002.

11: THE GENTLE VIRUS

Backhed, F., R. E. Ley, J. L. Sonnenburg, D. A. Peterson, and J. I. Gordon. "Host-Bacterial Mutualism in the Human Intestine." *Science* 307, no. 5717 (2005): 1915–20; doi:10.1126/science.1104816.

Battle, Y. L., B. C. Martin, J. H. Dorfman, and L. S. Miller. "Seasonality and Infectious Disease in Schizophrenia: The Birth Hypothesis Revisited." *Journal of Psychiatric Research* 33 (1999): 501–9; doi:10.1016/S0022-3956(99)00022-9.

Bézier, A., J. Herbinière, B. Lanzrein, and J.-M. Drezen. "Polydnavirus Hidden Face: The Genes Producing Virus Particles of Parasitic Wasps." *Journal of Invertebrate Pathology* 101, no. 3 (2009): 194–203; doi:10.1016/j.jip.2009.04.006.

Cawood, R., H. H. Chen, F. Carroll, M. Bazan-Peregrino, N. Van Rooijen, and L. W. Seymour. "Use of Tissue-Specific MicroRNA to Control Pathology of Wild-Type Adenovirus without Attenuation of Its Ability to Kill Cancer Cells." *PLoS Pathogens* e1000440 5, no. 5 (2009); doi:10.1371/journal.ppat.1000440.

De Filippo, C., et al. "Impact of Diet in Shaping Gut Microbiota Revealed by a Comparative Study in Children from Europe and Rural Africa." *Proceedings of the National Academy of Sciences* Early Edition (2010); doi:10.1073/pnas.1005963107.

Dingli, D., and M. A. Nowak. "Cancer Biology: Infectious Tumour Cells." *Nature* 443, no. 7107 (2006): 35–36; doi:10.1038/443035a.

Dürst, M., L. Gissmann, H. Ikenberg, and H. zur Hausen. "A Papillomavirus DNA from a Cervical Carcinoma and Its Prevalence in Cancer Biopsy Samples from Different Geographic Regions." *Proceedings of the National Academy of Sciences* 80 (1983): 3812–15.

Edson, K. M., et al. "Virus in a Parasitoid Wasp: Suppression of the Cellular Immune Response in the Parasitoid's Host." *Science* 211 (1981): 582–83; doi:10.1126/science.7455695.

Ewald, P. *Plague Time*. Garden City, N.J.: Anchor Books, 2002.

Ewald, P. W., and G. M. Cochran. "Chlamydia Pneumoniae and Cardiovascular Disease: An Evolutionary Perspective on Infectious Causation and Antibiotic Treatment." *Journal of Infectious Diseases* 181, Suppl. 3 (2000): S394–401; doi:10.1086/315602.

Flegr, J. "Effects of Toxoplasma on Human Behavior." *Schizophrenia Bulletin* 33, no. 3 (2007): 757–60; doi:10.1093/schbul/sb1074.

Hammill, A. M., J. Conner, and T. P. Cripe. "Review: Oncolytic Virotherapy Reaches Adolescence." *Pediatric Blood Cancer* 55 (2010): 1253–63; doi:10.1002/pbc.22724.

Kilbourn A. M., W. B. Karesh, N. D. Wolfe, E. J. Bosi, R. A. Cook, and M. Andau. "Health Evaluation of Free-Ranging and Semi-Captive Orangutans (*Pongo pygmaeus pygmaeus*) in Sabah, Malaysia." *Journal of Wildlife Diseases* 39 (2003): 73–83

Klein, S. L. "Parasite Manipulation of Host Behavior: Mechanisms, Ecology, and Future Directions." *Behavioural Processes* 68 (2005): 219–21; doi:10.1016/j.beproe.2004.07.009.

Lecuit, M., J. L. Sonnenburg, P. Cossart, and J. I. Gordon. "Functional Genomic Studies of the Intestinal Response to a Foodborne Enteropathogen in a Humanized Gnotobiotic Mouse Model." *Journal of Biological Chemistry* 282, no. 20 (2007): 15065–72; doi:10.1074/jbc.M610926200.

Lee, Y. K., and S. Mazmanian. "Has the Microbiota Played a Critical Role in the Evolution of the Adaptive Immune System?" *Science* 330, no. 6012 (2010): 1768–773; doi:10.1126/science.1195568.

Lo, S.-C., N. Pripuzova, B. Li, A. L. Komaroff, G.-C. Hung, R. Wang, and H. J. Alter. "Detection of MLV-related Virus Gene Sequences in Blood of Patients with Chronic Fatigue Syndrome and Healthy Blood Donors." *Proceedings of the National Academy of Sciences* Early Edition (2010); doi:10.1073/pnas.1006901107.

Muller, M., and L. Gissmann. "A Long Way: History of the Prophylactic Papillomavirus Vaccine." *Disease Markers* 23 (2007): 331–36.

*National Research Council. *The New Science of Metagenomics.* Washington, D.C.: National Academies, 2007.

Oriel, J. D. "Sex and Cervical Cancer." *Genitourin Medicine* 64, no. 2 (1988): 81–89.

Reddy, P. S., K. D. Burroughs, L. M. Hales, S. Ganesh, B. H. Jones, N. Idamakanti, C. Hay, S. S. Li, K. L. Skele, A.-J. Vasko, J. Yang, D. N. Watkins, C. M. Rudin, and P. L. Hallenbeck. "Seneca Valley Virus, a Systemically Deliverable Oncolytic Picornavirus, and the Treatment of Neuroendocrine Cancers." *Journal of the National Cancer Institute* 99, no. 21 (2007): 1623–33; doi:10.1093/jnci/djm198.

Reyes, A., M. Haynes, N. Hanson, F. E. Angly, A. C. Heath, F. Rohwer, and J. I. Gordon. "Viruses in the Faecal Microbiota of Monozygotic Twins and Their Mothers." *Nature* 466, no. 7304 (2010): 334–38; doi:10.1038/nature09199.

Riedel, S. "Edward Jenner and the History of Smallpox and Vaccination." *Baylor University Medical Center Proceedings* 18, no. 1 (2005): 21–25.

Rous, P. "A Sarcoma of the Fowl Transmissible by an Agent Separable from the Tumor Cells." *Journal of Experimental Medicine* 13, no. 4 (1911): 397–411.

Saarman, E. "How We Got the Controversial HPV Vaccine." *Discover,* May 17, 2007.

Schiffman, M., P. E. Castle, J. Jeronimo, A. C. Rodriguez, and S. Wacholder. "Human Papillomavirus and Cervical Cancer." *Lancet* 370, no. 9590 (2007): 890–907; doi:10.1016/S0140-6736(07)61416-0.

Sonnenburg, J. L. "Genetic Pot Luck." *Nature* 464, no. 8 (2010): 837–38; doi:10.1038/464837a.

Torrey, E. F., J. J. Bartko, Z.-R. Lun, and R. H. Yolken. "Antibodies to Toxoplasma Gondii in Patients with Schizophrenia: A Meta-Analysis." *Schizophrenia Bulletin* 33, no. 3 (2007): 729–36; doi:10.1093/schbul/sb1050.

Turnbaugh, P. J., et al. "A Core Gut Microbiome in Obese and Lean Twins." *Nature* 457, no. 7228 (2009): 480–84; doi:10.1038/nature07540.

Turnbaugh, P. J., V. K. Ridaura, J. J. Faith, F. E. Rey, R. Knight, and J. I. Gordon. "The Effect of Diet on the Human Gut Microbiome: A Metagenomic Analysis in Humanized Gnotobiotic Mice." *Science Translational Medicine* 1, no. 6 (2009); doi:10.1126/scitransl/med.3000322.

Turnbaugh, P. J., R. E. Ley, M. A. Mahowald, V. Magrini, E. R. Mardis, and J. I. Gordon. "An Obesity-Associated Gut Microbiome with Increased Capacity for Energy Harvest." *Nature* 444, no. 7122 (2006): 1027–31; doi:10.1038/nature05414.

Webster, J. P., and G. A. McConkey. "Toxoplasma Gondii-Altered Host Behaviour: Clues as to Mechanism of Action." *Folia Parasitologica* 57, no. 2 (2010): 95–104.

Widmer, G., A. M. Comeau, D. B. Furlong, D. F. Wirth, and J. L. Patterson. "Characterization of a RNA Virus from the Parasite Leishmania." *Proceedings of the National Academy of Sciences* 86, no. 15 (1989): 5979–82.

Williams, C. F., et al. "Persistent GB Virus C Infection and Survival in HIV-Infected Men." *New England Journal of Medicine* 350, no. 10 (2004): 981–90.

Yolken, R. H., F. B. Dickerson, and E. Fuller Torrey. "Toxoplasma and Schizophrenia." *Parasite Immunology* 31 (2009): 706–15; doi:10.1111/j.1365-3024.2009.01131.x.

Yolken, R. H., H. Karlsson, F. Yee, N. L. Johnston-Wilson, and E. F. Torrey. "Endogenous Retroviruses and Schizophrenia." *Brain Research Reviews* 31 (2000): 193–99; doi:10.1016/SO165-0173(99)00037-5.

zur Hausen, H. "Charles S. Mott Prize Papillomaviruses in Human Cancer." *Cancer* 59, no. 10 (1987): 1692–696; doi:10.1002/1097-0142(19870515)59:10<1692::AID-CNCR2820591003>3.0.CO;2-F.

——. "Condylomata Acuminata and Human Genital Cancer." *Cancer Research* 36 (1976): 794.

12: THE FINAL PLAGUE

Fair, J., E. Jentes, A. Inapogui, K. Kourouma, A. Goba, A. Bah, M. Tounkara, M. Coulibaly, R. F. Garry, and D. G. Bausch. "Lassa Virus-Infected Rodents in Regufee Camps in Guinea: A Looming Threat to Public Health in a Politically Unstable Region." *Vector-Borne and Zoonotic Diseases* 7, no. 2 (June 2007): 167–72; doi:10.1089/vbz.2006.0581.

Khan, S. H., et al. "New Opportunities for Field Research on the Pathogenesis and Treatment of Lassa Fever." *Antiviral Research* 78, no. 1 (2008): 103–15; doi:10.1016/j.antiviral.2007.11.003.

LeBreton, M., A. T. Prosser, U. Tamoufe, W. Sateren, E. Mpoudi-Ngole, J. L. D. Diffo, D. S. Burke, and N. D. Wolfe. "Healthy Hunting in Central Africa." *Animal Conservation* 9 (2006): 372–74; doi:10.1111/j.1469-1795.2006.00073.x.

Pike, B. L., K. E. Saylors, J. N. Fair, M. LeBreton, U. Tamoufe, C. F. Djoko, A. W. Rimoin, and N. D. Wolfe. "The Origin and Prevention of Pandemics." *Clinical Infectious Diseases* 50, no. 12 (2010): 1636–40; doi:10.1086/652860.

Russell C. A., et al. "The Global Circulation of Seasonal Influenza A (H3N2) Viruses." *Science* 32 (2008): 340–46; doi:10,1126/science.1154137.

Sharp, C. P., et al. "Widespread Infection with Homologues of Human Parvoviruses B19, PARV4, and Human Bocavirus of Chimpanzees and Gorillas in the Wild." *Journal of Virology* ePub (2010). January 28, 2010; doi:10.1128/JVI.01304-10.

Smith, D. J., A. S. Lapedes, J. C. de Jong, T. M. Bestebroer, G. F. Rimmelzwaan, A. D. M. E. Osterhaus, and R. A. M. Fouchier. "Mapping the Antigenic and Genetic Evolution of Influenza Virus." *Science* 305 (2004): 371–76; doi:10.1126/science.1097211.

Wolfe, N., L. Gunasekara, and Z. Bogue. "Crunching Digital Data Can Help the World-CNN." *CNN Opinion*. Cable News Network, February 2, 2011

Wolfe, N. "Epidemic Intelligence." *The Economist,* November 22, 2010, "The World in 2011," Science sec.

Wolfe, N. "All Risks Are Not Created Equal." Anderson Cooper 360. Cable News Network, April 30, 2010.

——. "Preventing the Next Pandemic." *Scientific American* (April 2009): 76–81.

ACKNOWLEDGMENTS

Because of the intense collaborative nature of the scientific research that these pages are based on, there are many people to thank. But before I do I'd like to express gratitude to those who helped me to make the book itself a reality. My outstanding agent, Max Brockman of Brockman and Company, Steve Rubin, president and publisher of Henry Holt and Company, and Paul Golob, the editorial director of Times Books, were all essential in this process. My editor, Serena Jones, provided keen direction and perspective along the way. Special thanks go to the outstanding students who have taken my Stanford University class, who helped through their thoughtful deep dives into the literature and wonderful questions. Finally to Kevin Kwan, who assisted with photos, and Robin Lee, whose diligence and daily assistance were invaluable.

As a scientist, my mentors who have shared with me their expertise and ideas over the years deserve credit for the positive elements of my work. Victor Jankowski gave me my first taste of the thrill of science as a child and Bill Durham at Stanford University helped to launch my research career. Richard Wrangham and Marc Hauser helped me to find the right direction for my work. Andy Spielman, of beloved memory, took me into his research group and under his wing and gave me the opportunity and training to conduct research that seemed odd to many at the time. Billy Karesh generously gave me my first opportunity to engage in field research and showed me how it was done. Don Burke, my postdoctoral mentor, has provided years of support and friendship and perhaps the greatest thing a young scientist can ask for: wonderful research projects and the freedom to pursue them. Don truly lives by his own quote: "You can accomplish almost anything if you are willing to give up credit for it." Debbi Birx, Larry Brilliant, Jared Diamond, Don Francis, Peggy Hamburg, Tom Monath, Ed Penhoet, Frank Rijsberman, Linda Rosenstock, and Jon Samet have provided much appreciated assistance and advice on my work as an independent scientist.

The work that formed the basis of this book would not have been possible without the collaboration of the forward thinking partners whom I have had the good fortune to engage with over the years. Early support by the Taplin Family to the Harvard School of Public Health, the National Institutes of Health (NIH) Fogarty International Center, and the US Military HIV Research Program helped to provide legs for my research. Ongoing partnerships with excellent programs in the US Department of Defense, including the Armed Forces Health Surveillance Center, the Defense Threat Reduction Agency, and the DoD HIV/AIDS Prevention Program have permitted the long-term continuity required to make real progress in this type of research, and the Henry M. Jackson Foundation for the Advancement of Military Medicine has played a key role in facilitating the process. Backing from innovative, generous organizations including Google.org, the Skoll Foundation, and the NIH Director's Pioneer Award program provided unique flexibility at exactly the right times and gave the potential for my research program to take pivotal new directions. Individuals within these and other organizations have shown incredible dedication toward moving the world in the right direction: Anna Barker, Debbi Birx, David Blazes, Larry Brilliant, Will Chapman, Dave Franz, Michael Grillo, Lakshmi Karan, Bruce Lowry, Nelson Michael, Sally Osberg, Jennifer Rubenstein, Kevin Russell, Toti Sanchez, Richard Shaffer, Mark Smolinksi, Joanne Stevens, Kofi Wurapa, and Cheryl Zook. Jeff Skoll, in particular has shown great vision and has helped to address the problem of pandemics with tools as varied as social entrepreneurship and feature film. The USAID Emerging Pandemic Threats Program under the leadership of Dennis Carroll now provides support aimed directly at one of my primary objectives—to predict and prevent the emergence of novel infectious agents. I feel fortunate to be able to participate in this important program with my exceptional collaborators at the University of California, Davis, the Ecohealth Alliance, the Wildlife Conservation Society, the Smithsonian, and the Academy for Educational Development. I thank Lorry Lokey for his generosity in funding the professorship in the Program in Human Biology at Stanford that I have the honor to hold.

I have had the good fortunc to collaborate with a wide range of first-rate scientists, clinical veterinarians, and physicians and their teams, many of whom have been mentioned in these pages, including Raul Andino, Francisco Ayala, Chris Beyrer, Patrick Blair, David Blazes, John Brownstein, Michael Callahan, Dennis Caroll, Jean Carr, Mary Carrington, Jinping Chen, Charles Chiu, Nicholas Christakis, Dale Clayton, William Collins, Robert A. Cook, Mike Cranfield, Derek Cummings, Peter Daszak, Eric Delaporte, Eric Delwart, Joe Derisi, Kathy Dimeo, Jon Epstein, Ananias Escalante, Jeremy Farrar, Homayoon Farzadegan, Jay Fishman, Yuri Fofanov, Tom Folks, Peter Fonjungo, Pierre Formenty, James Fowler, Pascal Gagneux, Alemnji George, Hillary Godwin, Tony Goldberg, Chris Golden, Jean-Paul Gonzalez, Greg Gray, Duane Gubler, Swati Gupta, Beatrice Hahn, W. D. Hamilton, Art Hapner, Kris Helgen, Walid Heneine, Lisa Hensley, Indira Hewlett, Tom Hughes, Warren Jones, Marcia

Kalish, Paul Kellam, Gustavo Kijak, Annelisa Kilbourn, Marm Kilpatrick, Neville Kisalu, Lisa Krain, Mark Kuniholm, Altaf Lal, Benhur Lee, Fabian Leendertz, Eric Leroy, Ian Lipkin, Jamie Lloyd-Smith, Chris Mast, Jonna Mazet, Wilfred Mbacham, Francine McCutchan, Angela Mclean, Herman Meyer, Matthew Miller, Steve Morse, Bill Moss, Suzan Murray, Lucy Ndip, Dianne Newman, Paul Newton, Chris Ockenhouse, Claire Panosian, Jonathan Patz, Martine Peeters, C. J. Peters, Rob Phillips, Brian Pike, Oliver Pybus, Shoukhat Qari, Steve Quake, Steve Rich, Annie Rimoin, Forest Rohwer, Ben Rosenthal, Kevin Russell, Maryellen Ruvolo, Robin Ryder, Warren Sateren, David Schnabel, Peter Simmonds, David Sintasath, Mark Slifka, Tom Smith, Joe Sodroski, Mike Steiper, Bill Switzer, Joe Tector, Sam R. Telford III, Judith Torimiro, Murray Trostle, Ajit Varki, Linfa Wang, Hugh Waters, Ana Weil, Kelly Welsh, Mark Woolhouse, Linda Wright, De Wu, Otto Yang, and Susan Zmicki. Adria Tassy Prosser played a pivotal role in helping to establish our program in Cameroon.

Special thanks go to my colleagues who make my life working internationally not only productive but a true pleasure: Ba Oumar Paulette, Dato Hasan Abdul Rahman, Mpoudi Ngole Eitel, Stephan Weise, Ke Changwen, Patrick Kayembe, Rose Leke, Jean-Jacques Muyembe, Koulla Shiro, Shuyi Zhang, Prime Mulembakani, Janet Cox, Balbir Singh, Edwin Bosi, Mahedi Patrick Andau, Olinga Jean-Pascal, and Emile Okitolonda. I thank them for welcoming me in their countries and homes. Appreciated also are the talented writers and journalists who have taken the time to visit us and who conduct the important work of translating science, like ours, to a broader audience, including Scott Z. Burns, Tom Clynes, Anderson Cooper, David Elisco, Sanjay Gupta, Anjali Nayar, Evan Ratliff, Michael Specter, and Vijay Vaitheeswaran.

I am incredibly fortunate to lead a team that includes some of the best and brightest at their work, who also manage to make my day-to-day life fun and exciting. Jeremy Alberga, Joseph Fair, and Lucky Gunasekara guide their respective teams to conduct the highest quality work under often challenging conditions. They impress me daily with their skill and dedication. Ubald Tamoufe and Alexis Boupda have been the best of partners in Central Africa. Mat LeBreton and Cyrille Djoko and their teams push science forward every day with talent and commitment. Karen Saylors, Corina Monagin, Erin Papworth, Maria Makuwa, and Kanya Long have all managed to conduct excellent work while managing complex programs in countries around the world. The work could not be done without the many other skilled scientists, technicians, and logisticians working at our headquarters in San Francisco or in labs and field sites around the world. I hope to work with all of them for many years to come.

I thank my father, Chuck Wolfe, mother, Carol Wittenberg, and sister, Julie Hirsch, and their families for always encouraging my love of science and standing decidedly by me despite my years of travel and absence, and my sweet grandmother, Ann Sloman, who has supported her grandson despite the fact that he didn't want to become a "doctor." I am privileged to have loyal and generous friends who have kindly contributed their valuable time and unique skills to my work, including Zack Bogue, Sebastian Buckup, June Cohen, Tom DeRosa, Jeffrey Epstein, Sanjay Gupta, Erez Kalir, John Kelley, Nina Khosla, Larry Kirshbaum, Boris Nikolic, Sally O'Brien, Sarah Schlesinger, Narry Singh, Linda Stone, and Riaz Valani. Last, but certainly not least, I would like to thank Lauren Gunderson, my thoughtful, wonderful reader, editor, and partner who kept me going throughout.

INDEX

Page numbers in *italics* refer to illustrations.

ABOUT THE AUTHOR

NATHAN WOLFE is the Lorry I. Lokey Visiting Professor in Human Biology at Stanford University and the founder and CEO of Global Viral Forecasting, an independent research institute devoted to early detection and control of epidemics. Wolfe received his bachelor's degree at Stanford and his doctorate in Immunology and Infectious Diseases from Harvard. He has held professorships at John Hopkins and UCLA, has published more than 80 articles and chapters and has been awarded research support totaling over $30 million in grants and contracts. Wolfe has been published in or profiled by *Nature*, *Science*, *The New York Times*, *The New Yorker*, *The Economist*, *Wired*, *Discover*, *Scientific American*, NPR, *Popular Science*, *Seed*, and *Forbes*. He was the recipient of a Fulbright fellowship in 1997 and the prestigious NIH Director's Pioneer Award in 2005. Wolfe has been elected a World Economic Forum Young Global Leader, a National Geographic Emerging Explorer, and in 2011 was named one of *Time* magazine's 100 Most Influential People in the World. He lives in San Francisco.